That One Patient

Ellen de Visser

That One Patient

Doctors and nurses' stories of the patients who changed their lives forever

Translated by Brent Annable

4th ESTATE · London

4th Estate
An imprint of HarperCollins*Publishers*
1 London Bridge Street
London SE1 9GF

www.4thEstate.co.uk

HarperCollins*Publishers*
1st Floor, Watemarque Building, Ringsend Road
Dublin 4, Ireland

First published in Great Britain in 2020 by 4th Estate
First published in the Netherlands as *Die ene patiënt* by Ambo\Anthos in 2019

2

Copyright © Ellen de Visser 2019
Translation copyright © Brent Annable 2020

Ellen de Visser asserts the moral right to be identified
as the author of this work in accordance with the
Copyright, Designs and Patents Act 1988

Nederlands **letterenfonds** This publication has been made possible with financial
dutch foundation **for literature** support from the Dutch Foundation for Literature.

A catalogue record for this book is available from the British Library

ISBN 978-0-00-837511-9 (hardback)
ISBN 978-0-00-837512-6 (trade paperback)

Typeset in Adobe Garamond Pro by Palimpsest Book Production Ltd,
Falkirk, Stirlingshire

Printed and bound in Great Britain by CPI Group (UK) Ltd, Croydon CR0 4YY

MIX
Paper from
responsible sources
FSC www.fsc.org **FSC™ C007454**

Foreword

My brother-in-law's funeral was held on a sunny afternoon in February, a day when he would ordinarily have jumped straight onto his ten-speed. Somewhere in that crowded room was a grieving oncologist. He'd taken the afternoon off, he told me, to come and bid farewell to a patient who had become his friend and from whom he had learned a great deal. His comment intrigued me: patients obviously learned from their doctors, since they are constantly explaining the origins of illnesses and what can be done about them. But the other way around? Were there perhaps other doctors roaming about, preoccupied with the memory of a particular patient, or with a story they will never forget?

That idea formed the inspiration for a series of columns published in the Dutch newspaper *De Volkskrant* in the summer of 2017: doctors talking about patients who had left a special mark on their lives and taught them valuable lessons. Conceived as a 'filler series' (the kind often used by papers to get through the quieter summer months), we initially only planned to publish six columns. Finding six doctors who were willing to come out and tell their personal stories would prove difficult enough – or so we thought.

The opposite proved true. Doctors were eager to participate and most interviewees knew exactly which patient they would talk about straight away. Our short-term experiment blossomed into a regular weekly column and after some time, doctors even took the initiative and began approaching us themselves. Very soon we cast the net wider to include not only doctors, but also nurses, psychologists, even a midwife and a medical social worker.

I rarely knew what to expect before each interview. I remember a greyish Monday morning in April, when a forensic pathologist drew me a sketch of the sun rising slowly above a field of wheat next to a bank by the road, where she had just confirmed the death of a young motorcyclist. Shortly afterwards I was back outside, dazed and slightly shell-shocked, solitary amid the hustle and bustle of the Amsterdam traffic as though nothing had happened.

Doctors and nurses need a unique form of professional empathy: at the same time as caring about their patients, they must keep their own feelings at bay as a barrier to shield them from the crushing emotional weight of their work. But there are patients who break through that barrier, who manage to touch and move their doctors in some way or other, ultimately shaping their thoughts and actions. And that is what healthcare professionals wish to share: stories confessing a certain vulnerability, one that continued to astound me, week after week.

In the medical world, emotions have long ceased to be a sign of weakness. In her book titled *What Doctors Feel*, American doctor Danielle Ofri explains how the quality of healthcare is highly impacted by doctors' emotions. Health professionals learn not only by practising the medical and technical aspects of their jobs, but also through their human experiences. In the words of one doctor: 'We have such intense contact with patients during a unique and often highly emotional period in their lives, it gives us food for thought, whether we like it or not.'

To our Dutch readers, it was as though these aloof, distant creatures in their white coats became more approachable with each passing week. They wrote in to say how moved they were by the doctors' openness, by the significance of the lessons they described and how they started looking forward to each new column. A poet once submitted an original piece for a psychiatrist; an old lady expressed her support for a young doctor in training who had made an error of judgement; a man admitted to bawling over his breakfast one Saturday morning, after reading an oncologist's story.

A few days after clinical ethicist Erwin Kompanje told of a young woman who had died in his hospital over twenty years earlier, I was contacted by the man who had been the patient's boyfriend at the time. He sent me a photograph of Irma, putting a face to the patient I had written about. I was granted a few more behind-the-scenes glimpses like this: while lying in a hospice during his final days, the patient of gastrointestinal specialist Joost Drenth read his own story when it appeared in the paper. He thanked the doctor for his courage, saying: 'Joost, you're a brave man. You never leave your patients empty-handed.'

This book tells the stories of those who gave their care providers the chance to grow and learn: to learn about their profession, about life and about themselves. For this English edition, I also wanted to include some new stories from British and American health workers, as well as stories about the COVID-19 pandemic, which continues to ravage the landscape of healthcare. Once again, I was surprised and thrilled by the overwhelming response and enthusiasm from interviewees. When I mentioned this fact to trauma surgeon Karim Brohi, he replied, 'Every doctor in the world has a story.' It's a privilege to be able to tell more of those stories in the final pages of this edition.

Ellen de Visser
Amsterdam, July 2020.

A Note on Euthanasia

Several of the doctors' stories in this book involve euthanasia. Unlike in Great Britain and the United States, euthanasia is legal in the Netherlands. Following an express request from the patient, Dutch doctors are permitted to provide assistance to end their life, subject to certain due-diligence requirements. These include an assessment of the patient's suffering, which must be deemed unbearable with no hope of recovery, with confirmation by a second independent doctor. Provided these criteria are met, doctors are not liable to be prosecuted. Four per cent of all deaths in the Netherlands occur by euthanasia; most of these patients are in the late stages of cancer.

The cyclist

Pieter van den Berg, oncologist

'I can still remember when the surgeon told me I would be treating a professional cyclist. Cycling is also a passion of mine, so I was looking forward to it and felt sure we'd get along. The surgeon added a disclaimer, saying that he was quite an eccentric patient. He wasn't wrong. Huib was distinctly unimpressed with doctors in general; he was straightforward to the point of bluntness, but with a sense of humour and irony that appealed to me. In our appointments we always ended up discussing cancer for one minute and cycling for twenty.

Ten months earlier he'd had surgery for colon cancer, and now it had metastasised. Chemotherapy was all we could offer him – a treatment that would extend his prognosis. On average, patients in Huib's situation had about a year left. But Huib scoffed at the statistics and continued to cycle massive distances week after week, scaling even the toughest mountains in Spain. He pedalled his way through chemo, saying that if it was going to hurt, it might as well hurt good and proper.

We gave him a year, but he lasted for over two. I can't prove that the cycling is what did it, but that's my hunch. It made him feel better overall and alleviated the side effects somewhat. His bike also brought distraction, allowing him to set the cancer aside for a little while.

There is no scientific proof that exercise helps cancer patients live longer. It may have more to do with general well-being, both mental and physical. Exercise strengthens the immune system, which in turn helps patients endure the suffering that comes with cancer treatment. I explain this to my patients on a daily basis and always cite Huib's case as an example.

Thanks to him, we are now considering a formalised exercise programme at the hospital. Why should we discourage patients from cycling to an appointment or treatment session? Our plan is to implement a buddy system, with volunteers who cycle along with patients, picking them up and dropping them off both to and from the hospital. I think it should even be possible for patients to cycle home after a chemo session – why not? Let's not underestimate their strength.

Over the course of two years, Huib and I became fast friends. I often wondered whether it was advisable, or even permissible – might there be a conflict of interests? But Huib always retained his autonomy. Our friendship deepened when I told him of my plans to open a chemo garden in the hospital, a pavilion where patients could relax and take their treatment in a natural, outdoor setting. Funding was still a major issue, so Huib suggested organising a sponsored cycling event. It was a huge success and together we raised fifty thousand euros. The path leading from the hospital to the garden is named after him.

I'm always very careful not to take my work home with me. Of course I feel for my patients, but when I shut the door behind me, those concerns need to stay on the other side. With Huib, that never worked. Never before had I grown so close to a patient.

When things started going downhill, he asked me to be the one to perform the euthanasia. I visited him at home to talk it over and he could see I was having a hard time with it. "Not chickening out, are you?" he asked, with a worried look. Then he hit back straight away with that biting humour of his: "You know who's having a hard time here? Me, that's who!" I knew I couldn't leave him in the lurch. But the afternoon when I ended his life was one of the saddest and most painful that I have ever known.

He once offered to give me a cycling jersey as a gift. I refused, thinking it wasn't proper for me to accept gifts from patients. After he died, his wife brought me a package. Inside there was a cycling outfit, with a note from Huib: *See? I get the last word after all.'*

Resilience

Elise van de Putte, paediatrician

'She was such a pretty girl, a dear little poppet, only two years old. The staff at the childcare centre had noticed bruises all over her body and a large blister on her foot. They called the domestic-safety hotline, who asked us to examine her for any potential signs of abuse. That's how the three of them came to be sitting in my office that day: the child, her mother and her stepfather.

We admitted the child straight away because of the bruises on her belly, which are sometimes an indication of internal injuries. We also wanted to be able to take our time and perform a thorough examination. It was my job to ascertain whether the girl's injuries were consistent with the story told by her parents. We X-rayed her entire skeleton – she had a break in her forearm and hairline fractures in several vertebrae. She'd been treated for a broken bone once before, when she had "fallen down the stairs". We hear that story a lot, but children fall over all the time, so it's entirely plausible. But those cracks in the vertebrae looked suspicious. They can sometimes appear spontaneously – due to naturally brittle bones, for example – but that wasn't the case with this girl.

We quickly got the impression that her injuries had been deliberately inflicted, but we still needed to do things by the book. It took us forever – I was so scared of coming to the wrong conclusion. I also felt really uncomfortable in the stepfather's presence; I found him intimidating. It was the little things that I found disturbing: the tone of his voice, the look in his eye. He mentioned in passing that he was a member of a shooting club. Everyone in the ward saw the girl stiffen whenever he entered the room ... we were in danger of making all kinds of unfounded assumptions and in cases like this, it's vital to keep your intuition out of it.

Her case showed me how incredibly precise I need to be in my job. It requires keen observation and for the facts of the matter to be established every step of the way. We asked national and international experts for advice: could this blister be caused by new shoes? Could this bruise have come from a fall? Because we were so thorough, I do feel as though I did everything I could for that girl. I say that, because we never found out who it was who had abused her. It's one of the most frustrating aspects of being a doctor or nurse here, but it's simply not up to us to identify the culprit. Our task is to try to support claims with facts: to demonstrate whether injuries are more likely to have been deliberate or accidental. The domestic violence organisation moved her to her grandparents' house for three months, where her mother and stepfather were allowed supervised visits. Everything seemed okay and she eventually went back home.

That little girl also made me realise just how insanely resilient and loyal children are. Offenders are often loved ones, which is what makes things so complex. She must have been in a lot of pain, but we never saw any sign of it as she really knew how to cover it up. She was so brave. We'd all grown so attached to her by the time she left, and now, whenever I teach doctors and students, I often tell her story. It helps me to process my own feelings too.

I still don't know what happened to that little girl. I'm not allowed to search for information, which I completely understand, but I find it frustrating regardless. I dread to be watching the news one day, hearing of a child who has fallen victim to domestic violence and then suddenly seeing her face. There are children whose stories keep me awake at night, and hers is one of them.'

Shakespeare

Erwin Kompanje, clinical ethicist

'It was in the early evening when I arrived at Irma's bedside: a young woman in her early thirties who had suddenly lost consciousness while out jogging one day. Scans had revealed a meningeal haemorrhage and the neurologist wanted to wait out the night. We kept her on the ventilator and monitored her blood pressure; any final decisions were put off until the next morning. But the outlook was bleak. She was in a deep coma, and the chances of brain death were high.

Her boyfriend was with her in the room. I was conducting my PhD research on brain death at the time, and was often in contact with patients' family members. I generally had no trouble keeping a professional distance, but this man broke through that barrier very quickly. Over the course of a long discussion, we connected on a deep level: he was an English teacher, and I'm a huge fan of English literature. That's what we talked about together, as night slowly fell.

I prepared him for the worst and told him that his girlfriend would probably die the following day. I quoted that famous line from *Romeo and Juliet* – "parting is such sweet sorrow" – as I knew that the sadness of his farewell would soon seep into all of his fond memories of their love and life together. He burst into tears.

That was when the imminent and irrevocable finality of his situation sank in, the realisation that this would be their last night together. He asked whether he might lie down on the floor in her room. I knew straight away what I had to do. We brought in an extra bed beside hers, dimmed the lights, turned down the equipment alarms. They lay next to one another: he put his arms around her, and together they spent their final night in peaceful tranquillity. I woke him the next morning at seven o'clock. A few hours later, the neurologist came to examine Irma one more time. It was confirmed; brain death was established and the ventilator was switched off.

Driving home that morning, it suddenly became clear to me just how much we take our lives for granted. Irma had gone out for a jog, thinking she would soon return home; her boyfriend had kissed her goodbye that morning, fully expecting to see her again soon. But clear skies can turn pitch-black in an instant.

He sent me a mourning card. I went to the funeral and was touched when he recited Shakespeare's words in his eulogy. Our final moment, or our last night with a loved one, will always come eventually. Usually we are spared the foresight of when that moment will be … but not him. He was grateful that I had been honest. The knowledge that Irma would probably not survive is what allowed him to decide how to spend his final hours with her. We were able to add to the sweetness of his sorrow, by creating the memory of their last night together.

Although it was twenty years ago, that night taught me just how important it is to value the simple things in life. Enjoying a cup of coffee with my wife, the comfort of sharing a bed, spending time with friends … happiness is in the human connections all around us. Life is an illusion of immortality. The farewells are coming, so create as many beautiful memories as you can.

Irma's boyfriend and I stayed in touch for a long time. He even came to my PhD conferral ceremony and to my wedding, five years after she was gone. Her tombstone bears the words that had touched him so deeply, that one line from *Romeo and Juliet* that I quoted at her bedside that night.'

The best option

Pieter van Eijsden, neurosurgeon

'She was six years old and had fallen from her bike one day without warning. I met her parents for the first time in the hospital corridor, when they had just heard the diagnosis: the MRI had revealed a malignant tumour on her brain stem. The prognosis was grim, as cancers growing from brain-stem tissue are incurable. The little girl did not have long to live.

Her father was resolute from the outset. If she was going to die soon anyway, then he and his wife refused to leave her in the hospital. They wanted to take her home, avoid throwing their lives into turmoil and instead enjoy the brief time they had left. That conversation stayed with me, and I often think back to it. Her life revolved around school, her father had said. She likes to play with her Perler beads and draw. If we take her home now, then she can keep on doing that for a bit longer. She'll never go to university, get a job, find a partner or chase her dreams, so why should we subject her to painful treatments that will only keep her alive for another few months?

Her parents were the best friends of my brother-in-law. After the diagnosis, he called me straight away to ask if I would help them through the period to come. Other doctors had given them several options: part of the tumour could be surgically removed, but it would always grow back again. Radiation therapy was another possibility, but would only postpone the inevitable. But one particular route – that of no treatment – was never discussed. After their initial response, the parents thought long and hard, and spoke to all kinds of experts, but ultimately stood their ground. They believed that for them, doing nothing was a well-considered choice and the best option for their daughter.

The little girl lived another seven months. During that time, the family concentrated on saying goodbye without the distraction or burden of invasive treatments and hospital visits. I learned a lot from those parents. I had always suspected that in some cases, doctors should tell patients that doing nothing is a perfectly reasonable option. But the longer I thought about it, the more uncertain I became. I personally didn't see the point in prolonging life unnecessarily, but was hesitant to express the thought to my fellow doctors. I also kept thinking: who am I to say? But suddenly there he was, this young father, so steadfast in refusing to try to save his most treasured possession. He echoed my own thoughts with pristine clarity.

Since then, I've become far less focused on treatment. For some patients, surgery is the obvious way forward; for others, treatment will clearly offer little benefit. But between these two extremes is an enormous grey area. Nowadays I ask patients what it is they still want from life. And to do that, I need to get to know them first.

It's already been a few years since that little girl died at home, and I spoke to her father again recently. Doctors, he said, tend to concentrate on the available treatments, while the unwanted side effects remain underemphasised. His experiences have changed the way I practise medicine. I now understand that a course of treatment can sometimes be too much, not only for the patient, but also for their families. Doctors find it very hard to say no, but sometimes it truly is the best option.'

One Tuesday night

Hans van Goudoever, paediatrician

'It was a Tuesday night and the gynaecologist had called me to the maternity ward to speak with a young couple. The woman was twenty-five weeks pregnant and the baby was already on its way. The outlook for children born so early is fairly bleak. At less than twenty-four weeks, treatment is usually fruitless; at twenty-six weeks we recommend it, but the two weeks in-between form a grey area. And so it was that, in the middle of the night, at the bedside of a woman in labour and with her husband beside her, I set about explaining exactly how likely it was that their child would survive and what the consequences might be, including the risk of any disabilities or permanent damage.

The parents told me of their career plans, of how they were planning to go abroad for work. They had no idea how a potentially disabled child might fit into the picture they envisaged and were also deeply concerned that their child might suffer. Eventually, they decided that it would be better not to offer any treatment once the child was born. Now, I am no stranger to these types of discussions and I can say that most parents beg us to do all we can to keep their child alive. I was stunned by the parents' response – nothing is more difficult than deciding against the life of your own child. But I was bound to respect their wishes, as all paediatricians and neonatal doctors have sworn to do. Their child, a little girl, was born in the early hours of the morning, and all we could do was try to make her as comfortable as possible. She died several hours later.

I thought to myself, I'll never see these people again. But one year later I got a telephone call from the same gynaecologist, with parents who wished to speak to me specifically. I recognised them straight away: it was the same couple. They told me that they hadn't moved abroad in the end, they'd had serious second thoughts. She had fallen pregnant again and was now already in labour after twenty-four weeks, one week earlier than before. They were having a boy.

I gave them the same speech about the baby's chances of survival, but this time their decision was different: they wanted us to do all we could to keep the baby alive. In the end, we couldn't: the little boy didn't make it.

The memory of their case has stayed with me throughout my career. We allow patients – parents, in this case – to participate in medical decision-making as much as possible. That's certainly important, as it's often said that people who can decide on their own treatment will always choose what's best for them. I now see what an illusion that is. These were well-educated people, faced with a gruelling dilemma and a ticking clock, with no friends or family to help them. In hindsight, perhaps they made the wrong choice. When I saw them again later, their heartache was palpable. I know I'm not allowed to change people's minds – I have to remain objective – but my experience with them has pushed me to make my message to patients even clearer.

Of course, we will never know what that little girl's fate might have been. Half of all premature babies never make it out of intensive care. Girls often do better than boys, which makes this case seem all the more tragic. If I'd been in their position that Tuesday night, I might have chosen differently. Isn't it always worth seeing whether a newborn has a chance at life? That little girl might not have made it in the end, but she was never even given that chance.'

Another world

Ben Crul, general practitioner

'She was an intelligent, engaging woman in her early forties, and my encounter with her strongly influenced the course of my life. I was a ward doctor, fresh out of university and on the path to becoming a specialist. She had been diagnosed with ovarian cancer, which often came with a grim prognosis. Her case was incurable. Every morning I went to see her with the nurses, but all I ever had for her was bad news. I happily visited other patients on my rounds, delivering the results of a scan or a blood test, but every time I walked into her room, it was with a heavy heart.

"Ben," she said one day (she always called me by my first name, that I still remember), "don't you ever have anything else to say? I know I'm going to die, but can't you talk about something else for a change, something nice?" Hearing those words, it hit me that patients are real people of flesh and blood, and that my job should be more than just doling out medical information. After that comment, I showed more of my personal side and began talking about other things – like my holidays, for example, she enjoyed hearing about that. Patients are more than just a disease – a fact that all doctors need to keep in mind.

After her question, it also dawned on me that I didn't really feel at home in the hospital. All around me specialists were scurrying about, busy with their next scan, their next round of chemotherapy … it was all so clinical. Sometimes, at the bedside of a patient who had been horribly disfigured by surgery, I would hear them talk of how marvellous it looked. I saw the look on the patient's face, as if to say: hang on, are you talking about me? I didn't feel at home there, and this patient helped me realise it. My conversations with her showed that specialisation was just not my thing and that I would much rather work in general practice.

So I started the GP programme and noticed the difference straight away. They told us it wasn't all just about the latest pathology results; it was about listening, looking, showing an interest. My mother died young, back when I was still at university, and I remember her doctor visiting us every second day. Although there was very little he could do, he always took off his jacket, sat down by her bed and was there for her. I never forgot how his mere presence gave so much support. Even then, I could see how important it was to take time for patients.

The lady with ovarian cancer died a year after I left the hospital. I didn't hear about it until a good friend of hers came to visit the practice where I was doing my foundation year. She had a bottle of wine with her and a letter written specially for me by the patient, shortly before she died. It was twenty-five years ago now, but I still get emotional thinking about it. She was so grateful for our brief time together and wanted to let me know that, even when she herself was no longer here. It just goes to show what patients and doctors can mean to each other – what I meant to her, sure, but also what she meant to me.'

The midst of grief

Hanneke Hagenaars, nurse

'He had started feeling unwell at work and just collapsed. They ventilated him in the ambulance on the way to hospital, and once admitted, his condition took a drastic turn for the worse. He was a young father to growing teenagers and had suffered a major brain haemorrhage. A few hours later the doctors confirmed that his brain activity had stopped completely. They spoke to his family, explaining that there was no hope of recovery and that any further treatment would be futile.

The doctors then consulted the national donor registry. There was no record of the patient in the database, so they very tentatively asked his wife whether the topic of organ donation had ever come up at home. He was still young and strong, and the doctors could see that many of his organs would be useful. His wife consented to donating everything that could be used – that was when my telephone rang.

I hurried to the ICU at the hospital, where I met the man's wife, sister and two children. They were uncommonly calm. I spoke to them at length, explaining in detail to them all – children included – precisely what the donation process would entail. In my head, I was constantly trying to figure out where they were at and predict what information they needed, which made the discussions quite intense. The doctors were performing tests to determine whether brain death had set in, a procedure that lasted several hours. When they were done, they came in to offer the family their condolences and told them the official time of death. I can still remember both that I wrote it down for them and what I said: "We'll take good care of him."

His wife stayed in the hospital during the surgery. Five hours she waited, until the last of his organs had been removed. Six weeks later I called her – something I always do, to tell people where their loved one has ended up – and she wanted me to pay her a visit. So suddenly there I was, standing in the unfamiliar sitting room of a relative stranger. And for the first time I saw a photograph of the patient when he was still alive, a man I had only ever seen lying immobile in a hospital bed, surrounded by tubes and equipment. His wife told me why she had not hesitated in giving her consent: they'd often talked about donation, about how it suited his personality and his outlook on life. He was always doing things for other people but had just never got around to putting his name on the register. When I told her that both of his kidneys, his heart, liver and his pancreas had allowed five people to lead normal lives again, she burst into tears.

I have dozens of these conversations a year, in hospitals throughout my region of the Netherlands, and they are always extremely intense. Occasionally one will stand out, a family member who sticks in my memory, and this woman was one of them. I only knew her briefly – during one of the most gruelling periods of her life, I suspect – but during those few weeks we became very close.

Later, when she recollected the moment I entered the room at intensive care, she said: "And then Hanneke came into our lives." That phrase says it all. I still marvel at the confidence and trust she placed in me, even when left alone with two children during one of the most trying times in her life. It made me realise why I find my work so worthwhile: I can offer a sense of consolation to those who are in the midst of grief, through the final gift that their loved one can give to the world. She helped me to see the importance of bringing a life to a fulfilling end.'

Bottlecaps

Marc Scheltinga, vascular surgeon

'He lived near Rotterdam and was in such terrible pain during a one-hour drive to Eindhoven that he had to stop twice to get out of the car and move around. That was how he hobbled into my surgery: doubled over in agony, exhausted and miserable. Doctor, he said, pointing at his groin, it feels like I constantly have six bottlecaps grinding into my insides. For three years his life had been dominated by a form of torment that other doctors said would never subside. The pain was debilitating; he was off work, living on a disability allowance. I was sitting opposite a man who had reached the end of his strength – if this is it, he said, then I'm done.

It had all started with a groin hernia. Initially the treatment seemed to have worked: the surgeon covered the rupture with a small piece of synthetic mesh, a plastic sheet used widely to repair hernias. The mesh is applied to the point of weakness in the groin, after which the body produces scar tissue around it, securing the mesh and closing the rupture. It's a reliable technique, and one that usually always works. Only now do we know that in extremely rare cases, the operation can have disastrous effects for the patient, causing chronic pain at the site of the operation. But when the man limped into my office that day, none of that was evident.

His own surgeon had ordered both an ultrasound and an MRI, both of which came back normal. They told him he would just need to learn to live with it and referred him to the pain management team. There he found little respite, as this kind of pain is hard to treat with drugs. The mesh needed to come out – that much he knew. He'd made enquiries everywhere, but nobody was willing to help.

That's when he came to our hospital. We are specialists in peritoneal pain, and his GP had read about us and written a referral. I examined him and concluded that he was probably right: the feeling of bottlecaps stabbing into his groin day in and day out was being caused by the mesh sheet. I called some of my fellow surgeons, but none of them had ever removed a piece of mesh before. Insertion was no problem, but removal? Nobody was game enough to touch it. So I thought, you know what? I'll just do it myself. I can still remember the moment when I told him: he broke down and cried. A grown man, big and strong, brought to his knees by a tiny piece of plastic and now reduced to tears at the prospect of meeting a doctor who finally believed him. It was that moment when I realised just how astutely doctors need to listen to their patients. Ninety-nine times out of a hundred, they are giving you the right diagnosis – you just need to put out the right antenna.

I had never performed the operation before. The mesh was completely overgrown with scar tissue, and snipping it free without causing additional damage wasn't easy. When he returned for his follow-up six weeks later, he was upright, cheerful and full of energy – a totally different person. His pain was gone, and he could enjoy life again.

His initial happiness soon gave way to frustration, however. Why had the whole process taken so long? Couldn't we have done it sooner? Why had three years of his life been wasted in agony? He was right, of course. He had become a victim of other people's indifference. If a doctor doesn't know what to do with a patient, then they have a responsibility to find somebody who does.

This rugged Rotterdammer taught me that we should always

be prepared to go off the beaten track, to do things we have never done before – especially if the patient will obviously be better off as a result. Provided you have good reason to do so, there is nothing wrong with going against the grain.'

Hide-and-seek

Erik Wehrens, tropical doctor-in-training

'His mother had brought him in, and he was already in a critical condition. When I arrived at his bedside, he was unconscious and suffering frequent epileptic fits. His breathing was deep and laboured. We quickly diagnosed his case: cerebral malaria, the most aggressive form of the disease. His chances of survival were slim. He was only six years old, and like so many children in South Sudan, would probably never reach adulthood.

The refugee camp where I work is bursting at the seams. Nearly 120,000 poor souls have been trapped here at the border town of Bentiu for years, driven from their homes by raging civil war. The conditions are dire: everybody lives in corrugated iron huts, and the camp is surrounded by swamplands – just the right conditions for the malaria mosquito. By far most of our patients are children. The hospital set up here by Doctors Without Borders has 150 beds, an operating theatre, an emergency department, a paediatric ward and a nutrition clinic. But despite the fact that we're doing our absolute best, we still see children die, week after week.

For four days I looked after this boy; my colleagues took over during the night. For those four days he lay unconscious in bed. We gave him anti-malarials, antibiotics and fluids, hoping to keep him out of danger just long enough so his body could repair the damage itself. The whole team knew of his case, we reported on his condition daily, but we all feared the worst. Even if he did survive, the likelihood of complications was high. Many children who survive such a virulent form of malaria become deaf, blind, paralysed or brain-damaged.

And then things started looking up. He responded to pain, then to speech, and twenty-four hours later his eyes flew open. He sat up and started eating and talking. Just over a week after arriving at our hospital, he was cheerfully running around again, playing hide-and-seek under the beds and sending footballs flying down the corridors. All of us were overjoyed to see him back to his old self.

Now, whenever I'm at a child's bedside and things are looking grim, I think back to that boy. Refugee camps are full of stories, many of which are sad and hopeless. But tales like this, of one boy's miraculous recovery, are what motivate us to keep going and not lose hope. The South-Sudanese doctors and nurses whom we support need it most of all: we do our six months and head home again, but they've been toiling here for years and see suffering day in and day out. It's so fulfilling to realise that we can sometimes make the difference between life and death.

The mother later told me that she had always assumed her son would make a full recovery. She'd brought him to hospital after all, hadn't she? Her attitude says a lot about the resilience of the people here. We're in the middle of a war zone, terrible battles are raging outside the camp, but most inhabitants maintain a positive outlook. Many suffer in silence, we see a lot of mental anguish, but people keep going and try to make the best of life. They build churches, hold markets, and there's football everywhere you look. They remain hopeful – even when things seem bleak – because they know what it is to live a life without hope.

I don't know what happened to that boy. He was happy and healthy when we discharged him, and there was no reason for him to come back. He has been swallowed up into the crowds of the refugee camp – somewhere out there is a little hut, where he continues to live as he did before. I hope for him that he's still running around, playing hide-and-seek and kicking a football about with his friends.'

Heart of stone

Wilco Peul, neurologist

'Peter was a nineteen-year-old student who had been brutally assaulted with a truncheon by strangers at the station one night. He arrived at the hospital in a deep coma. We operated on him successfully that night, and several more times afterwards. His prospects were grim, but we did all we could to keep him alive. And we succeeded: for months he lay in intensive care, and a year later he'd recovered enough to go home. He was so grateful, in fact, that he built a special website to tell others about his experience.

Peter returned to his home in Zeeland near the Dutch coast, where his parents lived, and I heard nothing more from him after that. I operate on dozens of patients a year with trauma-induced brain damage, often as the result of accident. I had never really looked into just how many are able to lead a fulfilling life afterwards. If patients left the hospital alive, that was the last I ever saw of them. They went back home, and if things didn't work out there, they were moved on to a rehabilitation centre or a nursing home. But I never saw the insides of those places.

Until four years ago, when I visited a rehabilitation centre for the first time and met people who might easily have been our own former patients. It was quite a confronting experience: I met people who were so severely disabled, who were capable of so very little, I wondered whether I would want that kind of life for myself. It got me thinking: what was, for me, the essence of humanity and happiness? Life becomes so fragile once the brain is damaged. I'd never considered whether we in the hospital were even doing the right thing. Should we always be doing all that we can, just because we can?

After that, the wheels in my head started turning. One of my students wanted to research whether patients who leave the hospital after brain surgery are able to lead happy and fulfilling lives, so we decided to go and interview our former patients from previous years. That's how we ended up in Zeeland, and I saw Peter again. We all assumed he was doing fine, as that was how we'd bid him farewell four years earlier. But he was now suffering frequent epileptic fits, his IQ had never returned to its former level and he was having memory problems. He had tried to study multiple times, but without success. And he was without a partner.

We were in serious shock, and for a full half-hour in the car on the way home we didn't utter a word. Thankfully we ended up in a traffic jam, which gave us the opportunity to talk things over calmly. We were so convinced that we'd given Peter a miracle, but right there under our noses, our miracle had turned into a mirage.

Through Peter, I now see the importance of the research we are doing now, in collaboration with other hospitals. How far should we really go when offering treatment? Should we always operate on every patient? Although our studies are far from over, by interviewing patients and their families we hope one day to offer well-founded answers to questions like these. That way, we can give better advice and information, perhaps even predict which patients will end up with a net positive (and which may not) and hopefully improve our decision-making processes.

My confrontation with Peter was life-changing. Not only has it helped to steer my professional decisions since, but it also altered my whole outlook on life. I realised I had become untouchable at work: I was invulnerable, impenetrable, and had been hiding my pesky emotions away where they couldn't bother me. For a long time, I had been working with a heart of stone – now, thanks to Peter, I have far more empathy to offer.'

The Underworld

Hans van Dam, nurse

'One Monday, on her husband's regular bridge night, she took an overdose. Her survival was an accident: there'd been a virus going around and some of the bridge players were sick at home. Her husband came home early and found her, which is how she ended up in the neurology department where I was the manager.

Doreen was a woman in her mid-thirties, with two young children. Although it was already thirty years ago now, I still see her face before me as if it were yesterday. She had a penetrating stare, but one that also seemed to look straight through you. All our attempts to make contact were rejected, she wasn't interested in talking. The attending neurologist said that he saw no signs of psychiatric illness, only of acute personal or life difficulties.

Little by little, something like trust built up between us. One day she asked to see me in private, so I took her to one of the smaller offices. She stared at the floor and said only how horribly miserable she felt. Long silences fell. Then suddenly I said: you know Doreen, I'm not here to stop you trying to commit suicide again.

That comment broke through like a wrecking ball. She looked up at me with eyes wide open and blurted: What did you just say? But isn't that all you people want? She was a smart woman and knew full well what would happen if she reiterated the desire to end her life. She would be institutionalised, put into compulsory care. So she held her tongue. I must have felt her insecurity and could sense that she would only open up once the pressure was off.

I told her doctor that I would keep in touch with her after she was discharged. As time passed, she started telling me about her life, a life that had brought her so much suffering. She had so little trust in others and was also hypersensitive and felt constantly bombarded with stimuli. Two years later she gave me Dostoevsky's *Letters from the Underworld*. It was a telling gift. Our communication was just like Dostoevsky's writings: she was revealing her own underworld to me.

Around four years later, it became clear to me that she wasn't doing too well. I had already suspected for some time that, sooner or later, she would make another suicide attempt. I asked her once whether she would and received a "maybe" as her response. The only person I told was her doctor, but he took no action, seeing the futility of it all. One morning I received a phone call from her husband: the previous night, Doreen had taken her own life in a horribly gruesome fashion. Her thirteen-year-old daughter was the one who found her.

My experiences with Doreen were a life lesson. Letting go of our presumptions – that's what we need to do. The medical profession is always so fixated on holding people back from the brink. Talk about arrogance! We are only ever guests in other people's lives, and that's how we ought to behave. Anybody in such desperate straits first and foremost wants to be understood. After that, it is our job to point out cautiously and clearly that help is available. That thought alone can give so much solace. I have learned to be a good listener; people without the chance to share their story become terminally lonely.

Doreen's horrific death has also influenced my opinion on assisted suicide. I believe that if a person genuinely feels they cannot go on, they should have access to a humane way out. That option alone can give people breathing space – either to soldier on, or to start moving towards a less solitary death. If the end is inevitable, things shouldn't end as they did with this young mother.

Her book is still on my shelf at home. The inscription in the front reads: *From Doreen.*'

Daughter

Piet Leroy, paediatrician

'It was Mother's Day, and my youngest daughter had fallen ill. At first it just seemed like a bad cold, but as the week progressed, she got steadily worse: a little girl not yet three years old, with a high fever, a cough and even shortness of breath. By Friday night she was in such a terrible state that I called my colleagues at A & E. She turned out to have acute pneumonia, there was pus in her lungs, even her blood pressure was abnormal – a pattern of symptoms that I should have recognised but had overlooked in my own child. I'm well-trained and had never missed a diagnosis – until then. Had it been anybody else's child, I would have gone searching for other symptoms to confirm my suspicions. But as a father, all I wanted was reassurance.

She was admitted to my own ward in paediatric intensive care. Suddenly I was no longer a doctor but a father. My perspective was literally turned around: instead of standing at the foot of the bed, I was now sitting beside it. The about-face taught me a lot. Doctors tend to take the disease itself as the starting point for their thoughts and communication: we establish the problem, formulate a prognosis and decide on how to fix it. Our goal is to provide as much information as possible. But the concerns of parents are entirely different. They're afraid, and they're wondering: will my child be alright? Will there be any lasting damage? Did I miss any signs? The best conversations I had as a father were with the doctors and nurses who not only gave me the right information but who also tested the emotional waters to see how I was feeling.

The experience left a major impact on me. I now take an entirely different approach when talking to parents. I know that my medical logic is often worlds away from what they are going through personally, a lesson that no book could ever have taught me. I take the time to get to know parents, speak to them about work, their hobbies, about family life. And I give them the opportunity to express their fears. That way we build up a relationship of trust, which opens the doors to more effective communication. Everything I say gets through much more easily after that.

My daughter had surgery several times. I was familiar with the condition and knew that it almost always worked out fine. That knowledge was what I was clinging to. But one week later, a serious complication arose: scans revealed a new infection between her lungs and her heart. That was the first time I saw true fear in the eyes of my colleagues. It lasted a day – a day when all of us were deathly afraid.

She made it in the end, but I have relived that fear dozens of times since then. In my mind, the room where my daughter had lain became inextricably linked to the drama of that one day: even after plenty of other children had come and gone, and I was back to being the ordinary paediatrician, that dread returned every time I entered the room.

When a child recovers from a serious illness, we generally respond with euphoria and tend to forget about the parents' gruelling emotions. Everyone feels relieved, the danger is averted, we're in the clear. Only now do I realise how long the sense of desperation can linger. The fear of losing a child, the sleepless nights, the hassles at home, the stress that parents all respond to in their own way . . . all of that can leave long-lasting after-effects. I now bring it up when I talk to parents: Things will be difficult for a while, I say, so take the time to process it all properly. My daughter's illness has enlightened me on the struggles of parents and made me a better doctor.'

Mother

Paula Groenendijk, nurse

'She enjoyed nightlife, fashion, holidays . . . an attractive and vibrant young woman in her late twenties, whose life had suddenly ground to a halt. She had late-stage cervical cancer, there was nothing more we could do for her other than try to relieve the pain. She was in my ward, and one evening she said to me: Paula, I can't go on like this. Her stomach and legs were swollen, she was in agony and utterly exhausted. I'm wasting away here, she said, when I should be in the prime of my life.

I usually do evening shifts, and the nice thing about that is that patients talk a whole lot more. Their visitors have gone home, the doctors have left, silence and darkness descend, and patients become reflective. I want euthanasia, she said. She brought it up again the next evening, and so I informed her doctor. He talked to her about it but explained that he wasn't prepared to take that step. There were still ways to relieve the pain, he said, and she still had a few good months left.

She was furious. There was a time when I would have understood her plight: I had long been convinced that euthanasia should always be honoured, even at a young age. I was in my early twenties and had just started out as a nurse when I assisted a doctor in the suicide of a young, terminally ill woman. None of my other colleagues were willing to help. I firmly believed that euthanasia was every patient's right, and I was always slightly outraged whenever doctors refused to do it. But now, after all those years, I was suddenly hit with crippling doubts.

Two months earlier, my son had suffered a heart attack and was treated in my own hospital. I sat beside him as I feared for his life. Now, at the bed of this young woman, I saw another mother in almost the same situation as I had been in two months earlier. My son and her daughter were the same age. My son had survived, but this mother was being forced to say her goodbyes. I could understand her daughter's desire to end everything – it was her life after all, it was her decision. I said as much to her mother, but for some reason it made me feel very uncomfortable. She got angry, saying it was far too soon for euthanasia, then asking me: "What if it was your own child?" That question really hit home, suddenly I could feel exactly what she was going through.

They eventually took the daughter home, where she died a few months later. Since then, there has been a change in the way I deal with young patients. It had always irritated me when family members told patients they needed to keep fighting. But not anymore. Since my son's emergency, I understand parents' fears and their resistance to euthanasia. It makes the final farewell so irrevocable. Now, I do all I can to keep young, terminally ill patients comfortable for as long as possible. I encourage them to leave their beds, and I treat them to good food and alleviate their pain as much as I can – all to prevent them from wanting to end it all too soon. I have come to understand the parents who aren't ready to face their child's death, as well as the doctors who feel unable to terminate the life of a young patient. Doctors are trained as healers, a conviction that clashes fiercely with the notion of performing euthanasia on a patient younger than their own children. Now I think that doctors also have a right to be unwilling – or unable – to end somebody's life.'

Cutting through the anger

Bart Fauser, gynaecologist

'They came to us not long after their diagnosis, a young couple whose future had started crumbling before their very eyes. The man had been diagnosed with cancer and was to undergo a course of chemotherapy that could potentially leave him infertile. He wanted to have some sperm cryogenically frozen so that he and his wife might still have a chance at children in the future.

Some time later they visited me again, with a concrete wish to start fertility treatment. Only a terrible dilemma had arisen: the man opposite me was now staring death in the face. The chemotherapy hadn't helped, and he was now terminally ill. Was honouring their wish the right thing to do? Would the wife be able to cope with the consequences of her decision? Were we justified in deliberately bringing into the world a child who would never know its father?

These were questions I could not answer alone, so I raised them with my monthly interdisciplinary consultation group, where these types of complex issues are discussed. Very quickly we reached the conclusion that it would be unwise to lend our assistance – at that time, at least. I remember the parents' response: they turned cold and distant. I explained how we as doctors feel jointly responsible for the treatments we offer and that we therefore have a say in deciding where the boundaries of our actions lie. I added that they were free to pursue treatment elsewhere, of course.

Three years later the wife was back in my consulting rooms, this time with her father. She told me that they had indeed found another clinic that was prepared to help them, but her husband had died during the first round of fertility treatment, and she had decided to stop. She'd been furious with me the first time, she said, but now she was back and ready to have her wish fulfilled. She had no new partner, led a stable life and had brought her father along to testify to it all.

This time we decided jointly to start treatment. After all, single women are perfectly eligible and her husband had given consent to use his sperm even after his death. I was moved by her strength in overcoming so much tragedy at such a young age.

Not long afterwards, I was contacted by the clinic where the IVF treatment had been scheduled. My patient hadn't turned up, so I gave her a call. I've been thinking things through, she said, and have decided not to go ahead after all. Three years ago, she went on, it seemed like your decision – now it feels like mine.

Only then did it hit me just how complex the decisions surrounding life and death can be. At first, she had seen our refusal as an imposition. Only later did she ask herself if her decision was truly what she wanted. It was a detour that she apparently needed to take in order to be sure of knowing her own mind.

I belong to the generation of doctors who have had to work out for themselves the best way to communicate with patients. With this couple, I think the line I initially took was far too hard. I had tried to help them understand our reasoning, but it seems I was unsuccessful.

By cutting through her anger and returning to me the way she did, this lady forced me into a confrontation with myself. I still stand by my initial decision, but she made me see just how important it is to build up a relationship of trust with patients first. Since then, I try even harder to involve parents in the decision-making process, so that the final decision is one that we can all feel good about.'

Escape

Sandra Bijl, general practitioner

'She had fled from Iraq, along with her husband and newborn son, who she had swaddled in a cloth and carried on her back across the mountains. After seeking asylum in the Netherlands, she came to live in my neighbourhood in Rotterdam, where she then had two more sons. But her hard-won safety was undermined by the drama on the home front. She came to my office complaining of vague symptoms, and only after numerous appointments did I finally uncover the underlying cause – she was being abused and raped by her husband at home and felt a deep sense of shame.

She told me what happened whenever he demanded sex against her will: she would let him have his way, pulling her headcloth over her eyes so she didn't have to look. She had fallen pregnant that way several times and had come to see me about an abortion without his knowledge. He became paranoid, convinced she was seeing someone else, and instructed his boys to keep watch on her all day long.

All the carers involved in her case were unanimous. You can't go on like this, they said, you have to leave him. But in the back of her mind was the shocking memory of her Iraqi brother-in-law, who had murdered his wife in an honour killing. She feared the same fate. She had called her three brothers – two who lived in Germany and one who lived in Iraq – to talk about a divorce.

Absolutely not, they said, unless she was prepared to move back to Iraq and leave her children behind. She was trapped.

Meanwhile, the pressure at home was rising. Her husband would casually tell his friends that she was a useless wife for refusing to spread her legs – in front of the children, no less, which added to her humiliation. She was also under pressure from other carers, who were threatening to take her children away due to the situation at home. All manner of organisations got involved with her family, and everybody thought they knew what was best for her. Even I put in my two cents. She was in my office on a weekly basis, and I urged her to take action, saying that there was no need to be afraid in the Netherlands. Her response was always the same: I can't. I even called Mayor Aboutaleb personally to see if he could reason with her husband, but I got no further than the secretary. We were all so eager to do her thinking for her, but only from within our familiar Dutch bubble. I neglected to consider that perhaps she might know best what the consequences of a divorce would be.

She ultimately found her own resolution. She had her brothers come over from Germany and arranged a family meeting with her husband. Initially her brothers thought she should persevere with the marriage, but she eventually convinced them otherwise and they came around to her side. On behalf of the family, they gave her permission to divorce, which allowed her honour to remain intact. She now lives by herself with her sons in the Netherlands, and her husband comes to visit the children now and again. He has had no choice but to accept the situation.

Her story taught me that I shouldn't think I can always solve other people's problems. Who am I to decide what's right for someone else? I work in a disadvantaged migrant area – what do I know of their culture and background? The strength of this woman astounded me: despite the incredible stress, pain and sheer exhaustion of her circumstances, she still somehow found the energy and courage to pull herself out of her misery. I hardly ever see her in my office anymore.'

Too nice

Mieke Kerkhof, gynaecologist

'Soon after the birth of her baby, I visited her at home to offer my congratulations. I go to see all of my new young mothers, but she didn't know that, and it made her feel so special. Little did I suspect. One day she handed me her baby's photo album – keep it for a while, she said, have a look through, write something nice inside. I should have realised how personal that gesture was, but I didn't see how I could refuse. She had a psychiatric condition and was one of the women we were treating in a special outpatients' clinic here at the hospital.

She had no partner, and her job started early, so when she returned to work I suggested that she come to my office personally for a follow-up – an offer that she interpreted as a kind of VIP treatment. We later found out that there were plenty of other things she had misinterpreted, but I ignored the signs, since it all happened so gradually.

She began writing me letters, leaving them for me at the hospital information desk. She told me that I had become a kind of second mother to her, and she wanted to meet up for tea to see whether I could perhaps play a bigger role in her life. That's where I drew the line: I consulted the hospital's legal expert and terminated the doctor-patient relationship via a formal letter. I wanted to refer her to a male colleague, but she refused.

That was when the flood of anonymous emails started, messages with a slightly erotic flavour, all from the same hotmail address: *shakespeare-in-love-to-be*. The sender seemed to know everything about me. It was unsettling – even intimidating. The emails continued for a year and only stopped once I threatened to lodge a police report. I never suspected that my ex-patient was the one responsible, until her psychiatrist asked me if I would be willing to have one last meeting with her. During that conversation, she confessed that she had been stalking me the whole time.

She was transferred to another hospital, and that was when I made a crucial error, one that taught me a valuable lesson. In an effort to spare the new female doctor the same fate, I decided to try to warn her. Doing so required me to violate the rules of patient-doctor confidentiality, but I was convinced that it was necessary, that I couldn't simply withhold my knowledge. I called her office, but there was no answer and I was put through to her secretary, who I urged to refer the patient to a male gynaecologist. The secretary – quite unfortunately, I'll admit – called the patient directly to discuss the matter. My ex-patient was livid and lodged seven official complaints against me with the hospital, making a tirade of accusations. They were all dismissed in the end, except the complaint about violating patient-doctor confidentiality – that one, they ruled, was justified.

I now see just how vulnerable we are as doctors. It's an intimate profession as it is, gynaecology even more so. This woman made me feel as though I had abused that intimacy. Of course, her behaviour was most likely due to her condition – she had a compulsive personality disorder – but I can't ignore my own part in the drama. I'm too nice, too eager to do all I can for patients. My colleagues often say that I get too close, that I keep too little distance. I find it hard not to, it's just how I'm put together. Still, since my experiences with this patient, I've become much stricter on myself.

Because of the valuable lesson it contains, I tell this story to all of my new doctors in training. Be dedicated to your patients, by all means – but also be sure to set your boundaries.'

Medical logic

Hester Oldenburg, breast-cancer surgeon

'I did what I'd seen so many patients do – I ignored it. But when I felt that little lump in my right breast growing slowly, I knew it was bad news. It took me another several months to approach one of my colleagues about it. He felt the area, did an ultrasound and took a biopsy. I clearly remember getting the phone call from the pathologist; the news travelled through the ward like wildfire. We're a close-knit team, and it was unnerving to know that we were just as vulnerable to breast cancer as our patients were.

I wanted the surgery in my own hospital, of that I was certain. I felt such a bond with my fellow surgeons, where else could I possibly be any better off? The hardest part was having such a detailed knowledge of the procedure myself. It was a surreal morning when, for the first time, I went to work as a patient rather than a doctor. Being wheeled into the operating theatre, all I saw were familiar faces.

The operation was followed by five weeks of radiotherapy, a treatment whose effects I had grossly misjudged – another thing I saw my patients do all the time. Breast cancer is common among middle-aged women, who are often juggling so many balls: work, family, ageing parents. They think they can fit the radiotherapy in among their other commitments, and so did I. The reality caught me completely by surprise. As I felt the wound tighten, all I could think was: this is hurting me, make it stop. As a doctor I was used to making assessments, evaluating outcomes and reviewing statistics, but suddenly all my well-trained medical logic was out the window. The surgeon and radiotherapist sat down with me, had a talk and got my feet back on the ground. I finished off the radiotherapy, and two weeks later I was back at work. But then, seeing every patient sitting opposite me was like looking into a mirror. All those people were me; their stories were far too close to home. I attended a congress in Berlin, where hundreds of doctors from all over the world met to discuss breast cancer, and I couldn't stop thinking: these people have no clue whatsoever. After that I took three months off. If I wanted to be a doctor again, I needed to stop being a patient first.

I was re-diagnosed with breast cancer six years later during a follow-up scan. It was in the same breast, buried so deep I couldn't feel it. After radiotherapy, it is no longer possible to preserve the breast during a second surgery. I would need a full amputation and reconstruction, which meant five days in hospital, drains, a catheter . . . the first day after the operation, I couldn't even sit up on the edge of the bed.

I talk a lot to patients about the period that awaits once the treatments are over, when everyone thinks life will just return to normal. But it's only then that you finally realise what has happened. And worse – what could have happened. That eerie, oppressive feeling, I underestimated it both times and now I also see that doctors ignore it completely when dealing with their patients.

I almost never tell patients that I've had breast cancer myself; the consultation is about them, after all, not me. But the experience has changed me as a doctor. I now know first-hand what women face, since I've been through it all myself: the sense of weakness, the fear, the anxiety.

But most of all, I now see just how pervasive and persistent cancer is. In my case, I knew it wouldn't kill me, but it was still a wake-up call. I realised for the first time that I can't take my health for granted, that cancer can happen to me too, that I can't escape my mortality. I'm for ever a different person now.'

In the mirror

Irene Mathijssen, plastic surgeon

'She was seven years old when I saw her for the first time: Katie had been born with severe skull and facial abnormalities. Her cranial bones had fused prematurely before her birth and impeded the growth of her brain. Her eye sockets were shallow and far apart, and her upper jaw was pushed back, which hampered her breathing. She'd already been through multiple surgeries and had come to us for a full facial reconstruction. We were to pull the entire centre of her face forward. It's a major procedure: we separate the bones of the forehead, eye sockets and jaw, twist the two halves of the face towards each other and then pull it all out slowly using a special apparatus.

The operation was a success. Look how pretty you are now, we all said, and her parents told her that she now looked more like her sisters. It was all very well meant, everybody was happy – except Katie. In the mirror she saw a completely different face; she didn't recognise herself anymore. Everyone started treating her differently. Katie is no fool and figured out exactly what was going on. Although she was precisely the same girl, the people around her had become instantly friendlier and more enthusiastic. And all because she now looked like other children? Like she "should"? All of a sudden – and it's quite sinister, when you think about it – the world now also believed she was more intelligent. Katie knew it and felt deeply distressed.

She tried to keep her frustration hidden, but eventually her mother noticed how she was struggling with other people's opinions and came to us for help. The social worker and psychologist from our team spoke to Katie and worked through her feelings. They sorted it out in the end but to us, her tale was a real eye-opener.

We now have very different conversations with the children who pass through our operating theatre. We realise that, even at such a young age, they are conscious of their appearance and aware of its effects on the outside world. Now we tell children that we operate on them so that they can close their eyes better, for example, or to make breathing or chewing easier. The fact that they might look better afterwards is something we never say anymore. We ask them beforehand what they think things will be like after the surgery, the psychologist is there, and we talk about Katie's story. We explain that they will look very different afterwards and ask: how do you feel about that?

I'm still in touch with Katie. She once sent me a copy of a talk that she gave to her class, which was a touching account of her condition, her hospital visits and a list of all the things she can't do – forward rolls, for example. She's thirteen now. I recently got an email from her, saying that folks still stare at her a lot wherever she goes, that many people find her different, or weird-looking. She spoke about it to one of her friends, who suggested maybe having a minor tweak – just a small operation, so that her appearance won't change too drastically all at once, to avoid a repeat of her previous experience. I'll be operating on her again soon.

The temptation is great to tell children that surgery will make them look more beautiful or normal, but those are all value judgements. It took a little girl of seven to teach me that, and it's a lesson I draw on with every single patient I see.'

War

Selma Mogendorff, general practitioner

'He was the kind of patient I hardly ever saw. A big, strong fellow, nothing ever wrong, a construction worker who only ever booked an appointment if he'd sprained his knee or hurt his hand somehow. He was full of optimism, even a little cheeky, with a presence you couldn't ignore. Everything will come good in the end, he said, and if not, well, then it's just your time to go. He worked in his own business until he was at least seventy. I used to ask him whether the hard labour wasn't too taxing or if he was scared up there on the scaffolding – "Course not, love," he'd reply. He brought me flowers now and again, since he felt like he never gave me much business.

He once told me, fairly casually, that he'd been stationed in Germany as a young man during the Second World War. I never got the impression that he was bursting to tell me about it, so I didn't pursue it. That is, until he fell seriously ill with dementia and I began making regular house calls. In his head, he had returned to his youth, and none of his stored memories as a young man could be repressed any longer. There he sat, in his chair by the window, a frail eighty-year-old, and all he could do was wail and scream with a kind of primal fear. "God, oh God, no, it can't be!" he cried, "these aren't people no more, they're just corpses ... the humanity ... oh boys, what'll we do? What'll we do?"

Who knows what horrors he witnessed. Perhaps he worked in

the factories, with others from the concentration camps? Although it was fifteen years ago, I still remember that I could sometimes hear his whimpering as soon as I stepped through the front door. It was heart-wrenching, there was so much pain and sadness coming to the surface. And guilt, I think. He had obviously wanted to do something – but what? What could he have done, as a young lad of eighteen?

It was too late for lucid conversation. I got no coherent answers to the questions I asked, he just kept repeating the same thing over and over, about how horrible it all was and how much anguish he suffered. If only I'd been given a chance to reason through his past, to help process his experiences with him, perhaps I could have brought him some peace. But now he was unreachable. It was like he went off to war every day, I saw it happen and there was nothing I could do. I felt so powerless.

His wife told me he had never said a word to her about it, not even when they'd been on holiday to Germany. For all those years he had kept his past bottled up, hidden away beneath a jovial exterior. That was his survival strategy. But now he was powerless. He could no longer keep the memories at bay, and they came flooding back.

If I had seen but a small sign of his distress, I would have made some enquiries. I suspect that for a very long time, he hadn't been ready to open up about his experiences in the war. Now I see that patients don't always show their true colours, that their behaviour can sometimes mask strong emotions. Unpleasant or irritating behaviour, too, can be a sign of anxiety or grief. Since my experience with him, I pay greater attention to how people act, and take the time to ask myself: why are they behaving like this?

He died a few months later and, to be honest, I was happy for him. The final stage of his life must have been terrifying. All those memories, formed so long ago, but suddenly so real . . . when I try to imagine it, it breaks my heart.'

Shadows

Peter de Leeuw, physician

'It was a December night in 1977, and I had the weekend shift. I was a third-year foundation doctor with lots of patients under my charge, both in my own ward and elsewhere. In those days, it was still normal for the trainee doctors to work for three days and nights straight, and I'd already been going for twenty-four hours at least. What happened must have been a combination of my lack of experience and total exhaustion.

She was around sixty years old and had been in hospital for a while with a vague set of symptoms. Her blood acidity was elevated, but I couldn't figure out why. I ordered an X-ray of her abdomen, and I remember when the nurse came to show it to me. It was late in the evening, in a half-lit room, I was tired and had just got off an unpleasant personal phone call – so my head was a little frazzled. I looked at the X-ray and saw nothing untoward.

She died that same night. I was called over, and at first I was puzzled – the X-ray had been normal, after all. An autopsy was performed, and the cause of death turned out to be a stomach perforation. I should have alerted the operating theatre, so she could have been operated on immediately.

We retrieved the X-ray to have another look. A hole in the stomach always produces a pocket of air in the abdomen, which should have been visible on the X-ray. And on a second inspection, there it was – the faintest of shadows, but there nonetheless. I had made an irrevocable error. Today, I have enough knowledge and experience to know that acidosis is frequently a sign of gastro-intestinal problems. Of course, I should have rung up the physician on call. But as an assistant in those days, you always thought twice before calling up a man at home and dragging him out of bed like that. Us young doctors had to work things out for ourselves, more or less.

The matter was dropped. The unwritten rule back then was that doctors don't make mistakes, and the events of that night were never discussed any further. I was left to process it alone. I didn't dare raise it with any of my colleagues, I was too ashamed of what had happened.

For a long time, I was petrified of slipping up again. I hardly dared to treat anyone during the period immediately afterwards, and when I spent some time in the outpatients' clinic, I made sure to go over every X-ray at least twice. It took a while before I regained my self-confidence. I became a lot more cautious at work, and for many years my personal life took a back seat. Once I had qualified as a doctor of internal medicine myself and was on call at home, I would frequently head out to the hospital after somebody phoned, just in case. Whatever was going on at home was always less important. I subconsciously recalled my own unpleasant phone call from that night, I think, which had thrown me off balance and perhaps contributed to my fatal blunder.

Now I know that all doctors make mistakes, and my own experiences have made me more lenient when training new doctors. I am not so quick to judge. A mistake is easy to make, and young doctors can be destroyed if you come down on them too aggressively.

That case was forty years ago now. Time has had its healing effect, but I still regularly think back to her. The events of that night have become an apparition – one that has haunted me ever since. There are many patients I will never forget, but this woman stands out the most. It was my fault that things ended badly, and that feeling has coloured the rest of my life.'

A man's voice

Idie Pijnenburg, nurse

'He'd been a resident in my department for a couple of years, an incredibly withdrawn sixty-year-old, unable to cope with stress and incapable of expressing himself. His pent-up frustrations would often erupt in aggressive outbursts, and occasionally even into psychotic episodes. His wife had become afraid of him and wanted him out of the house. We all got along with him just fine. He had his own room and enjoyed working in the garden. His wife often visited at the weekend, or sometimes he would go to see her back home. He was fully aware of his biggest problem and formulated it in his own special way: "I can't talk," he'd stammer, sullenly shaking his head.

One day he got a sore throat, which turned out to be a tumour on his vocal cords. That this very man – who himself said that he was unable to talk – should develop a laryngeal tumour was astoundingly symbolic. He needed an operation, but one that would leave him voiceless forever. His vocal cords needed to come out along with the tumour; the only alternative was to slowly choke to death. We tried to talk it over with him, but he became panic-stricken and fell into stress-induced psychosis. His wife, who had a mild intellectual disability, was helpless and didn't know what to do. She asked us to make a decision for them. So our team consulted with the doctors, and together we looked at the best course of treatment.

The surgery took place in a hospital far away, where he knew nobody and where his family could seldom visit. How isolated and alone he must have felt. The operation, however, was a success, and he returned to us without his voice. We were all prepared for the fallout from a trauma from which he would psychologically never recover. But to our great surprise, the opposite occurred: he slowly made his way back into the world. He used gestures and even started writing things down to get his thoughts across, something I'd never seen him do before. He learned to talk again by pressing a small device against his larynx. Teaching him to use it wasn't easy, but he was so determined that he sometimes even tripped over his own words. Over the course of that first year, he became a totally different person – cheerful, engaged and optimistic.

Many of his old tensions had disappeared, and we formed a theory as to why. Once all the pressure and expectation to speak had been removed, he no longer felt inadequate. For the first time in his life, he could relax during a conversation. If he said anything at all, it was a bonus, and he even received compliments on his words. The operation that saved his life also brought him a sense of mental freedom, an outcome we never could have predicted. His surprise recovery taught me a valuable lesson, one that I've carried with me for the past thirty-five years and that I put to use in both my professional and private life: keep an open outlook, entertain the notion that things might be different to how you think. We are often so sure of our beliefs, but our convictions – whatever they are – can be mistaken. At the time, as a psychiatric nurse, I thought I had people pegged, that I knew enough to figure them out. With the best of intentions, I often claimed to know what was best for them. But since that one patient, I am no longer one hundred per cent certain of any conclusion.

I kept in touch with him after I left that organisation, occasionally visiting him and when he died, I attended his funeral. During his final years he always looked far happier than he'd ever done before he lost his voice. And he never suffered a psychotic episode again.'

Grandchild

Alex Gosselt, intensive-care foundation doctor

'Six weeks of intensive care had taken its toll. His VAD was infected, he'd been through multiple surgeries and was on heavy antibiotics, but still his condition worsened despite all our efforts. His kidneys were failing, his respiratory muscles had atrophied due to long-term ventilation and the most recent X-ray showed that the abscesses in his chest were only getting bigger. At the start of my evening shift, the team met to discuss what we might still do for him. He wouldn't survive another surgery, that much we agreed on. We were forced to conclude that in his case, we were out of options.

We broke the news to him and his family. He was conscious but a little hazy, so we weren't sure whether he'd properly understood. He said he wanted to return home, and we were willing to help make that happen. Later that night, as I was walking past his room, one of the nurses stopped me to ask if we might use the intensive-care ultrasound machine to do a pregnancy scan. I was hesitant; evening shifts are often busy and unpredictable, I would probably have no time and the IC machine was unsuitable anyway. Which of the nurses is pregnant, I asked? Her answer surprised me: it wasn't a nurse, but the daughter of the extremely sick man. She was seventeen weeks pregnant and would so very much like for her father to see his grandchild before he died. My doubts evaporated, and I quickly resolved to find out what could be done.

I recalled that the internal medicine ward had a serviceable ultrasound kit. I called them up, and they said I could borrow it. There was just one small problem: I'd never performed an ultrasound on a pregnant woman before. I told the nurse that she should go and prepare the family and perhaps manage their expectations a little, then began looking up instruction videos on YouTube.

Somewhat timidly, I stepped into the man's overcrowded room. The mood was downcast. I said that I would do my best, but that it was my first time and I didn't know if I would succeed in getting a clear picture on the screen. We shifted the man across to the wall and wheeled in a second bed for his daughter; I was positioned between them, and we tilted the monitor to give the father a better view. My hands were trembling, but no sooner did I place the probe on his daughter's belly than, to my surprise, the foetus appeared: first a waving arm, then the baby's beating heart.

The atmosphere changed instantly; the mood became uplifted, and the family's enthusiasm was palpable. I tried to remain calm, but of course I shared in their happiness. The man stayed quiet at first, but his response, when it came, was marvellous: new life comes, he said, where old life departs. It seemed that he knew what he was facing after all and understood that he didn't have long to live. He left the hospital a few days later and died at home.

What a special privilege it was to have the chance to make this man and his family so happy after hearing such tragic news. And with so little effort from our side. During life's final days, a small gesture can often be of incalculable worth. In our profession, where we continually walk the fine line between life and death, I now see even more clearly how that experience can be one of the most beautiful of them all.'

Clear pictures

Huub Buijssen, psychologist

'The scenes he had witnessed were seared onto his retina: people trapped in burning cars, the panicked cries of dozens of wounded victims. He and his motorbike-patrol partner had sped to an accident on the A16 not far from Breda, where thick mists had caused a massive pile-up on the motorway.

Two days after the accident, I got a phone call about the officer: he wasn't doing well, and they wondered if I, as a military psychologist, might be able to help. When I saw him, he was still in a state of shock. Though he had always been mentally very stable, now he was worried that he would never return to normal, that he might never wear his uniform again. He wondered why he and his partner were having such a hard time, when virtually the entire force had witnessed the accident. They'd all seen the same thing, hadn't they?

I let him talk it out. It was my first ever "emergency" consultation, and on the way there I'd thought about how to best explain the mechanics of trauma processing. I eventually decided to draw him a kind of graph, made of five parallel horizontal lines. The top line, I began, represents ecstatic happiness, the bottom line is the deepest imaginable mental distress, and the centre line is a normal state of mind. Then I added a wavy timeline shaped like a concertina. I explained that after a traumatic experience, we always dive straight down to the very bottom. Our world is in chaos, and everything looks black. Not too long afterwards, we experience moments of distraction that allow us to clamber upwards for a stretch – until we are reminded of the trauma, and we crash back down again. But never as far as before; after that we rise back up, a little higher than last time. Psychological processing is a waveform, I told him, which slowly but surely gravitates back towards a balanced state of mind.

A month after our session I asked how he was doing. He told me he felt much better and had been back at work for three weeks. I was curious about what had given him the most support. I'd always thought that sharing one's story, putting words to the tears, was the most effective part of any consultation. But no – to my surprise he said it was my drawing that had helped him get back on top of things. It had helped him understand what was going on in his head and offered some reassurance and perspective. I brought normality to his abnormality and turned him from a patient into a person, a human being with a normal response to a harrowing experience.

Although the accident was over twenty-seven years ago now, that officer taught me an important lesson: after a trauma, it's essential to show people the course that their healing will take. Later, when I began training doctors and nurses how to offer support to their colleagues, I heard over and over and again just how valuable a clear explanation can be.

The officer was also relieved, he told me, once he realised that his response hadn't been extreme. On the day of the accident, he and his partner thought they were riding to the scene of a minor collision and were totally unprepared for the horrific scene they encountered. I impressed upon him that through his actions, he'd successfully spared his colleagues the same emotional shock: the detailed information he relayed to the radio room gave all other incoming agents the chance to mentally steel themselves for what awaited them. That fact gave him some consolation, letting him know that his suffering had not been for nothing. Since then, I always try to present the victims' behaviour during the trauma itself in a positive light. This officer showed me just how important that can be – how something so seemingly worthless can suddenly seem worthwhile after all.'

A carefree child

Nens Coebergh, confidential reporting doctor

'She was a girl of six and was at the doctor's quite a lot. Her school had contacted us, because according to the mother, her daughter was constantly ill. The reports became ever more dramatic: first it was gastrointestinal problems, then shortness of breath and fainting spells. The mother sent her to school with medication that the teacher was to administer if ever she suffered an attack. The strange thing was that at school the girl was absolutely fine. She was carefree, cheerful and played happily with the other children. The principal said that to them, she didn't seem sick at all.

The mother's GP sympathised with her and offered his full support. The girl was referred to an academic hospital, where she was thoroughly examined. I tried to tell the hospital doctors the principal's side of the story, that she exhibited no symptoms while at school. I urged them to base any diagnosis exclusively on their own observations. If you ask parents how their child is, you are trusting them to give you an honest answer. Things might be different here, I thought. But they wouldn't listen. I can still remember how isolated I felt. How could I impress upon them that there might be more to the story? The girl went from specialist to specialist and was subjected to invasive procedures that can only have been horrible for such a young child. The doctors came up empty-handed. I had no way of reaching the mother. This all happened a long time ago, and procedures were different back then: we never spoke to the parents, only to professionals.

Shortly before the school holidays, the school called to tell me she was scheduled for heart surgery. I was shocked, but relieved when I heard that the hospital first wanted to keep her in for observation, to see what the problem was. After the holidays, it turned out the mother had cancelled the hospital stay. Things had suddenly improved: her daughter saw the doctor less and less, and she told the staff at school that she thought her daughter might just grow out of it after all. I felt somewhat reassured, but my concerns didn't subside completely.

Then the school called again with terrible news: the little girl had died. She'd phoned her grandfather in a panic one afternoon, because her mother had become aggressive. Granddad went over straight away but was too late – the little girl lay dead at the bottom of the stairs. Only then did I hear that a few weeks before all of the mother's dogs had been found dead in their kennels. The entire neighbourhood knew about it and had felt for the family – who on earth would do such a horrible thing? If I'd known sooner, I probably would have been on my guard straight away, as there is a proven connection between animal cruelty and child abuse. During the ensuing court case, it came to light that the mother suffered from Munchausen syndrome by proxy and had deliberately been making her daughter ill to attract attention. It was she who had killed her daughter and poisoned the dogs.

I will never forget that girl; the events surrounding her case have shaped me as a confidential reporting doctor. Is there anything more I could have done? Could I have saved her? Would our more modern procedures have prevented the tragedy? I just don't know. Though I voiced my suspicions at every turn, I was all but isolated in my distrust. Be vigilant, stand firm, verify everybody's testimony and check anything that might be important – that is the sad lesson I have learned. That mother duped the lot of us, which cost the lives both of her pets and her daughter. The very thought of it pains me to this day.'

Future

Paul van Zuijlen, plastic surgeon

'One hour before midnight, a devastating fire broke out in a Bucharest nightclub. It was a Friday evening in October, and the line-up included a rock band who used pyrotechnics as part of their show. The stage caught fire, smoke filled the room and hundreds of panicked concertgoers frantically clambered over each other to reach the narrow exit. The local hospitals couldn't cope with the numbers of wounded flooding in and called for international aid. One week after the fire, the Ministry of Foreign Affairs asked the three Dutch burns units to take on some of the victims, and that's how a young woman from the Romanian jet set ended up with us in Beverwijk.

She was in shock when she arrived, and in a critical condition. Her body and face were covered in third-degree burns. Our primary concern, however, was the bacteria she had brought in with her: it was an aggressive strain, we'd never seen anything quite like it, and she was immediately put into quarantine. Our antibiotics did nothing, and covering the wounds in artificial skin would only have created a better environment for the bacteria. All we could do was continually flush out the wounds and transplant real pieces of skin little by little. The strategy worked, and we were gradually able to combat the virulent bacteria.

Initially we kept her under sedation, so I didn't get to know her until weeks later, when we finally gained the upper hand and she was slowly woken up. With a fierce gaze, she asked me the same question over and over: doctor, what will I look like? She had seen her bandaged hands straight away of course and knew that I had been forced to amputate all her fingers. But her face – what would happen to her face? She was anxious, kept demanding a mirror. I explained that she needed to be patient, that scars take a long time to heal. I can still recall the conversations at her bedside. I want to get better, she said, I want to be beautiful again.

I was dreading her reaction. How would such an attractive, stylish woman from the fashionable elite live a new life with facial disfigurements? I was convinced that her fate would bring her down. But no: instead I saw an incredible source of strength come to the surface. She looked herself squarely in the mirror – then had the courage to show her face to the world. She returned to her old life and started posting pictures of herself on Instagram, dressed up to the nines with bare arms and exposed midriff. It was inspiring to see how she not only accepted but also took control of her situation.

That's what makes my job so rewarding: seeing how people can bounce back after a terrible accident and how that inner strength sets the course for the rest of their lives. People who get themselves moving, literally and figuratively, have a crucial head start. Burns heal better and more cleanly in patients with get-up-and-go. But more importantly, a healthier recovery awaits those who have the courage to accept and own their injuries. This young woman is proof of that.

Her story is always in the back of my mind. I now take heed of people's outlook on life and pay closer attention to their mental fortitude. My job is more than just surgery – the psychological side can be just as important, so I encourage my patients to look ahead and focus on the future.

Now, more than two years after the fire, I see photographs of a strong woman with a radiant, dazzling expression. Nobody sees her scars or fingerless hands anymore – it's her eyes that get all the attention.'

Christmas

Soufian el Bouazati, A & E doctor

'It was Christmas, and my shift had just started, when one of the nurses came to warn me that a baby had been brought in and wasn't doing well. Three days earlier the mother had been to the emergency GP, because her baby was having trouble breathing and wouldn't drink properly. Now they were both in A & E. When I walked in, the paediatrician had already started his examination.

Half an hour later, the nurse's fears were confirmed: the monitor showed the baby's heart rate declining rapidly. I began massaging the baby's heart, taking the chest in my hands and pressing on the sternum with my thumbs. The X-ray we had taken showed an enlarged heart; the baby was in the middle of heart failure. We tried drugs to get the heart started again; the paediatrician was on the phone with the intensive-care specialist from the neigh-bouring academic hospital, who was on his way over in an ambulance. My hands were getting sore, but there was no time to think about that. Our entire team was clustered around the child; the tension throughout the group was palpable.

Half an hour later, the specialist arrived. We tried different drugs to try to get the heart beating again. My hands were cramping up, and I could feel the baby's tiny chest becoming steadily more rigid. And the whole time, the baby's face was right there in front of me. When I couldn't go on any longer, one of my colleagues took over the resuscitation. The paediatrician had gone to talk to the mother.

Shortly after that, the specialist looked towards us and took stock of the situation. He came to the conclusion that there was nothing more we could do and asked whether we all agreed to cease the resuscitation. We had tried everything we could, but to no avail. The baby died right there, in our very hands.

Then came the mother's response: she'd been sitting in the corner of the room the whole time and had seen everything. She pushed us aside and took up the child in her arms. That was when I broke down. I had a baby too, the same age as hers – they even looked alike. I realised it could just as easily have been my own child lying there. I locked myself up in the coffee room and let go of my emotions – a flood of feelings that I didn't know what else to do with. After all, doctors don't cry . . . do they?

We all felt defeated. But there wasn't time – other patients needed our care. At the end of that shift we had another resuscitation case, an elderly lady aged eighty-two. Her breathing was troubled, and on the monitor, her heartbeat suddenly stalled. We couldn't save her either. But I was emotionally numb by that stage. The contrast in my feelings that evening was enormous.

I hardly slept and instead lay awake wondering if there was more we could have done, or something different. I awoke with a start early the next morning, with the baby's face before me and the events of the previous night replaying in my mind.

Our job is emotionally demanding; we experience a lot of tragedy and can't carry it all with us all the time. To maintain a lucid, rational state of mind, we shut ourselves off. That strategy works fine for a while, until things get too close to home – that's when the dam breaks and the emotional barrier gives way, and that's what happened to me that night.

The following year I had the Christmas shift again. A seventy-year-old man was brought in, and we resuscitated him successfully. But my thoughts still strayed back to that little child . . . since then, no Christmas has ever been the same.'

Plastic bag

Dick Bisschop, gynaecologist

'Early one morning, I received a phone call from the A & E nurse. A fifteen-year-old girl had come in with severe abdominal pain and pain while urinating. The GP had diagnosed a bladder infection, but when her symptoms didn't subside, her stepfather had brought her to the hospital. The nurse – in her wisdom and experience – smelled a rat. This is no bladder infection, she said.

I took an ultrasound of her belly and saw immediately what the problem was. The girl had an abnormally enlarged uterus, which meant that she must have given birth very recently. I asked her point blank where the baby was, but she seemed incredulous. I remember being quite terse in my response. I turned to the stepfather. "Go home and look for a baby," I said.

At that moment, the girl's mother came in carrying a plastic bag. She had found a dead baby outside the house beneath the window sill, with the umbilical cord still attached. Her daughter had given birth to a child alone that night and had presumably panicked. The girl was unresponsive, so I couldn't find out whether she'd even known she was pregnant beforehand. The autopsy showed that the baby had been born alive but hadn't survived the freezing overnight conditions. Everyone in the ward was in shock. Since I couldn't state natural causes on the baby's death certificate, the public prosecutor was informed and an investigation followed.

We kept the girl in for observation over the next few days, and the nurses tried to get her talking. I wanted to know just how traumatic the previous night had been for her. We never found out. I saw her several times more during consultations, when I also observed her mother's recriminations. They came from a big South American family with lots of children, one extra would have been no problem, she said. Why had she let the baby die?

The nurses told me later that they frequently saw the girl wandering through the hospital afterwards, roaming the corridors near the nurseries during visiting hours. We didn't really know what to do . . . she must have been in quite a state.

If nothing else, the drama surrounding this girl has given me a completely new perspective on abortion. It all transpired during the late eighties; the Abortion Act had just been passed, but I still fiercely opposed the termination of any pregnancy. I was brought up in a southern Catholic family with eleven children, and we'd all been instilled with a great sense of respect for unborn life. But this girl's story confronted me with a reality that my studies had never shown me, a world where the problems are so great that abortion was often the best solution. Later, after working for some time in Africa and the Caribbean, I saw many young girls whose pregnancies brought them nothing but trouble. I could close my eyes to it no longer, and since then I have done a lot of work to promote contraception among young people.

It wasn't until much later that I started wondering whether I'd truly done enough for this girl. Had I fallen short, perhaps not medically, but emotionally? My children were the same age – perhaps I responded too much like an angry father? I called the principal of her school and asked him to keep an eye on her for me. Should I have asked after her, to find out how she was doing? The family had quickly shut everybody out and blocked almost all contact with the outside world. But despite all the barriers, now, when I look back, I do feel a sense of regret.'

Bucket list

Annelies van Vuren, trainee physician

'She really didn't belong in our ward, a seventeen-year-old girl surrounded by older adults. But the children's hospital had too little experience with her condition – how often do you see a teenager with a melanoma? Her skin cancer had metastasised, and she'd just started immunotherapy. She was in pain and couldn't stay at home anymore. That's how she ended up in our ward. We put her in a single room, which turned into a typical teenager's pigsty before we knew it.

She had problems taking her medication, so we put her on an IV drip. It fed directly into an artery just above the heart, and it's standard procedure to take an X-ray of the lungs to make sure the needle is properly inserted. The image of her scan is forever burned into my memory: I have never been so shocked in my life. Lungs are supposed to appear black on X-rays, but hers were lit up like a Christmas tree. Cancer everywhere, not a corner was spared. We had all so hoped that the immunotherapy would help, but this image, this explosion of tumours, dashed all our hopes instantly to pieces.

It was a Friday afternoon when I gave her the news. I sat on a stool beside her, she was cross-legged on the bed, telephone in hand. It was a tough blow, I had to take all her hope away. The metastases were not only in her lungs – they'd spread to her abdomen as well and perhaps even her brain.

I had no clue how long she had, but it certainly wasn't long. I asked her what she still wanted from life, and how we could help organise it for her. In the days that followed, those were the questions that dominated the ward. She said she wanted to marry her boyfriend. So together with the family, we began organising a wedding in the hospital sanctuary. It was just like the real thing: she had a dress, there was a cake, and even speeches. And despite being the source of all her misery, she wanted me at her wedding. She was a radiant figure, a fluffy cloud of white tulle in a wheelchair. After the wedding came the second item on her bucket list: she recorded an episode with her favourite YouTube vlogger.

Her situation couldn't be all fun and games, of course, and we also had more serious conversations. There was the subject of strong pain medication, for example, and her fears of gaining weight due to the stimulants we had prescribed for her. She wanted to discuss euthanasia, but that idea was soon eclipsed by all her other wild plans. Gradually, I saw her condition decline. And as her bouquet was still drying in the linen closet, our talks turned more and more towards the very end. She wanted to go home, and so one last time we turned the ward upside down to make it happen. A few days later, she died.

Her story has stayed with me to this day. She brought a tornado of youthful adolescence to the department and in those final days left behind a crazy legacy. That girl gave us a one-week crash course in how to live life.

I hope that we sent her and her family off with some fond memories. I once told her story in front of hundreds of medical specialists. She was always fond of attention – well, that's what she got. I also received responses from many of my colleagues: her story spurs us on every day to give everything we can, they say, to do our utmost for patients, even though we sometimes feel powerless. We would have so liked to cure her, but it was out of our hands. Hopefully the day will come when we no longer need a seventeen-year-old girl to remind us of our mortality.'

Living from love

Anne Marie Alders, paediatrician

'Pearl was around six years old when, running around in the backyard one summer's day, she very briefly lost consciousness. That had already happened once before, during her final school swimming exam. I suspected heart arrhythmia, but the hospital tests produced no diagnosis. Their theory was hyperventilation, so they gave her some breathing exercises to do, which seemed to help.

Two years later, I received an unsettling telephone call from my mother. Something's happened, she said, there are ambulances all around the swimming pool. It was a Monday morning in August, and I feared the worst. I drove straight home, and my suspicions were confirmed. It was the first day after the holidays, the weather was warm, and Pearl had gone to the pool with some friends after school. Once in the water, her heart stopped. The ambulance was on the wrong side of the pool and had to circle all the way around, and in those days there were no portable defibrillators anywhere. It was a long time before they got her heart beating again. Too long.

Two days later, an MRI scan showed that she'd suffered serious brain damage, with no chance of recovery. Pearl had turned out to have congenital heart arrhythmia after all, and the physical exertion at the pool combined with the shock of the cold water had proven fatal. For weeks she lay in intensive care, until the doctors and her parents eventually decided that further medical treatment was futile. But when the ventilator was switched off, something miraculous occurred: Pearl kept breathing, and could even swallow too – precisely the bodily functions she needed to stay alive. She celebrated her eighth birthday in intensive care, totally oblivious to her situation.

Pearl is my neighbour, and her parents are my friends. That Monday was fifteen years ago, and they've been caring for her ever since. They even built a special extension to the house just for her. Of course, they receive help from a special team of dedicated nurses, but the day-to-day care is mostly up to them. I am now Pearl's doctor, and along with her GP, I'm on call to help whenever she needs medical attention.

I see the love they shower on their only child together. I see how well they manage, but also what an incredible struggle it is, and I have so much admiration for them. Pearl is included in everything they do. Wherever they go, she goes. She spends time outdoors every day, she goes with them on holidays, her birthday is always a big celebration. Her parents have found the strength to pick up where they left off and to give meaning to her life, which at one point seemed entirely hopeless.

Pearl's story has changed me as a doctor and as a person. I've started looking at life from many different perspectives and now realise far more clearly just how vulnerable parents are, especially parents with chronically ill children. No matter how much help they receive, once the front door closes, they're on their own, and that must feel so lonely. I think about that aspect so much more now. I listen and try to offer all the support I can, so they don't feel isolated.

Pearl is now twenty-three, and for her, time seems to have stood still. All around them, her parents see how differently things could have turned out. Pearl's friends from that day at the pool are all out studying, having fun, starting their first relationships. But Pearl is stuck at home. She'll never go on a date, never take a boyfriend anywhere. Day after day, her parents toil for her unconditionally and receive so little in return. They themselves always use the same touching words to describe their situation: Pearl is living from love, they say.'

Humour

Marcellino Bogers, nurse

'I was eight years old when my mother fell seriously ill. She had skin cancer. Pain medication wasn't as advanced back then, and I remember waking up often at night to her cries of pain – a devastating sound to a child. The next morning, when I got up, I would always play the clown, putting my pyjama pants on my head to try to make mum laugh. Because if mum laughed, then all was well. I was too young to realise then that she was going to die.

Ten years later, I went into nursing on a work-study programme. My first placement was in geriatrics and on my first shift I helped to look after a dying old lady. The shift manager was supposed to supervise me, but he got called away. The lady died that night, all alone, with only me beside her. I was barely eighteen, and it hit hard. At home after my shift, perched on the edge of my bed in my student flat, I decided I should quit the programme. It was clearly too much for me; there was no way I'd make it through.

The next day I went back to work anyway. I was allocated to a room with eight patients and was given the task of washing a gentleman who was clearly in a foul mood. What's the matter, I asked, did we start the day on the wrong foot or something? At that point he threw back the covers, and to my astonishment, I saw that one of his legs had been amputated. I was mortified – how could I have just blurted it out like that? But the man just let out a huge belly laugh. The tension was dispelled, and the humour of the situation chased away the unpleasant memories of the previous night.

From that point on, humour became my survival strategy. I was convinced I would survive the profession, if only I could lighten up the serious moments with a little fun. As time went on, I discovered that patients, too, are really in need of a good laugh. After my unintentional quip, I eventually became good friends with that man in the nursing home. Humour can help to bridge the gap between carers and patients, and shared laughter can strengthen the bonds of trust, offer relief, put worries and cares in perspective and serve as a stepping-stone towards more difficult conversations.

I started experimenting. Cautiously to begin with – you do need to test out whether humour is something that patients appreciate, and if so, what kind. And though I never joke about sensitive subjects, I've learned that there's no reason why a serious illness should stand in the way of a good joke. Patients don't want their lives to be all doom and gloom, and a little humour can help cut through a sombre mood. It's the terminally ill patients I've laughed hardest with – for them, laughter can briefly eclipse their fear of death. Getting patients to see the humour in their own situation can also make them feel like they still have some control over their lives.

I've made it my mission to introduce humour into healthcare. I give lectures and workshops and have published a book. But there's no manual to follow, and obviously you should not play the clown the whole time – you need to develop a sense for it. But there's always something to laugh about with patients. The situations present themselves, you just need to learn how to recognise them.

Only now do I see that my motivation stems from the little boy I once was. At the very least, my aim is to coax a smile out of patients – it was my mother who taught me just how valuable a smile can be.'

Motorcyclist

Bertine Spooren, forensic pathologist

'My pager went off very early one morning. I was on call, so I rang in to the communications centre – they needed to send out a pathologist. Shortly beforehand the police had been notified of a body discovered down by a dike, probably a road accident. I was to help the police with their examination of the scene. I drove down in my car to an address in the polder, a large expanse of land reclaimed from the sea. The morning roads were still empty.

Polderlands are vast and flat, and the roads are elevated above dikes that slope downwards to the fields. I spotted the police car from some distance away. Parked next to it was the traffic accident examination van, while at the base of the dike lay the unlucky motorcyclist. The shattered fragments of his helmet were scattered some distance away. He must have lain there all night, invisible to the cars driving past. He'd only been spotted that morning by chance, by a farmer perched high up in his tractor who happened to glance down to the right.

I examined the body for rigor mortis, to determine when death had set in. I checked for lividity, or bruise-like pools of blood in the parts of the body lying closest to the ground. I looked at his position and surroundings – the only sand on his body was under his fingers, so nobody had moved him around. He had died some-time during the previous evening or night, I concluded, after falling off his bike and smashing to the ground below with incredible force.

The police officers clambered back up to the van to get a blanket to cover him up with. We called a mortician. I stayed down below, along with the body and my colleague from the traffic accident department.

Just at that moment, the sun came up. We were standing at the edge of a field of wheat: a light mist hung in the air, the birds started chirping and the sideways light lent the landscape an almost dreamlike quality. My colleague saw the look on my face. "Breathtaking, isn't it?" he said. We were surrounded by the beauty of nature, a vision of glory and serenity, while beside us a young man lay dead. I thought of his family, perhaps a wife and children, who in all likelihood were anxious because he hadn't returned home that night. I knew that the police were on their way to bring the news of his death. I knew he would have died instantly, and although I really should always remain objective, I do remember finding it a comforting thought. I've examined plenty of bodies, and usually their deaths seem quite distant to me. I've learned to keep my barriers up and avoid getting too emotional about what I see. But on that sunlit morning, something was different. The magnificence of the surroundings transported me away from the situation, while at the same time drawing me further into it. I was suddenly struck by how utterly unpredictable life is, by the realisation that it is completely up to chance whether we live to see another day. This is how things can go, I thought: it was pleasant weather, this man went out for a ride on his bike ... and just never came back. And he died somewhere out of sight, with the traffic passing by, invisible to the world. It hit me hard, and I found it somehow symbolic of the indifference of the universe. A man dies, but the world keeps turning, and the sun rises once again.

Despite the tragedy, I didn't find it a troubling thought. On the contrary, the sun emerging above the fields imparted a sense of consolation: death is such a central part of our existence, it can exist in the very midst of life.'

Village chief

Sjoerd Zwart, general practitioner

'It was around lunchtime when the hospital message boy cycled up to my house. My placement at the rural hospital in Ghana had only just started. I'd been on duty the night before and needed some rest, but I was still available for emergencies. The messenger handed me a note from a nurse: a lady aged around thirty had come in with stomach pains. Her symptoms didn't sound serious, so I wrote back saying I would be there in an hour.

The messenger came back ten minutes later with another letter. It said to come immediately, as the patient was a relative of the village chief and a member of the Ashanti tribe, the most powerful in Ghana. *Please, doctor,* he urged me in English. I felt slightly annoyed and wondered why this woman should feel so entitled to special treatment. I sent a reply with the same message as before, but it didn't really sit right with me, so after half an hour I went to the hospital anyway. My suspicions about her condition were correct: the woman had a stomach flu, so I gave her some pills and sent her home.

Two days later, a delegation from the village chief appeared unexpectedly before my front door. The chief had been insulted, they said, because I had declined his urgent request. I first spoke in my defence, saying that the urgency had been unwarranted. But I also apologised to them in the end, since their presence made me realise the importance of conforming to the norms of the place where I was a guest. Like it or not, their society is simply more hierarchical, and those of higher status enjoy priority treatment. I felt humbled. There I was, the high-and-mighty western doctor, the white fellow who thought he could teach the locals a thing or two.

That patient in rural Ghana taught me a lesson that I would think back to time and again. Once back in the Netherlands as a GP, I started noticing how often I was confronted with the very same dilemma. Special treatment is considered inappropriate in our society, but doctors do it all the time. I regularly ask hospital specialists if one of my patients can skip the queue, and there's a good reason why I keep emergency slots open at the beginning and end of my usual consultation hours.

Priority is often given for medical reasons: the more serious cases are dealt with first, of course. But in Ghana I learned that sociocultural aspects can also factor in. My patients with a Turkish background, for example, often panic whenever a child has a fever, because meningitis is very common in Turkey. Their anxiety is usually unfounded, but I always quickly make time to see them whenever they call. I would do the same for a sick professor who was on their way to an important conference. If David Beckham needs surgery after a major injury, or whenever the queen needs a doctor, they also take priority, right? Nobody even thinks twice about it. Presumably it's because society still benefits, in some indirect way. No two people are truly equal, that's what became clear to me during my time in Ghana.

I was worried that I would lose my job after the incident, since the chief also turned out to be president of the hospital board. But we talked it through together afterwards and not another word was said. When my daughter was born one year later, I named her after the queen mother, the most powerful Ashanti woman in the district.'

Grumpy old men

Wilco Achterberg, geriatric specialist

'His medical file didn't even fit inside the folder: a thick wad of paperwork written by dozens of specialists, a compendium of all his physical ailments. The cause of his conditions was still unknown, and he had eventually just wound up in a nursing home. He was in his early seventies and in so much pain that all he could do was lie still in bed. Nobody knew what was wrong with him. He was angry, frustrated, ill-tempered, difficult to deal with and constantly fighting with the nurses.

We consulted the psychiatrist, who made a very unorthodox suggestion. He called it the "paradoxical approach": just listen, he said, give plenty of attention and resist the urge to try to solve or explain every symptom straight away. I was to see him for an hour once a week, only to talk, without reverting instantly to "doctor mode".

From then on, I saw him for an hour every Friday morning at nine o'clock. I would sit at his bedside and simply let him talk. Initially he was always quite agitated, but would calm down as the conversation went on. During those first few months, my meetings with him felt like a real chore. I was inexperienced, still in training and was dying just to be a doctor, to get out there and treat people. With him I felt like I was treading water, playing a part, waiting it out. Until I noticed that our conversations were actually doing him good. He began to look forward to our visits, as they made his whole week seem easier. I got a dressing-down from him whenever I was late – and a talking-to from the staff if I skipped a week, since he would always revert to his snarky ways.

I came to realise how much he needed my attention, and it's a lesson I've carried with me throughout my career. Elderly people are often contending with a wide range of conditions, and as doctors we sometimes make the mistake of sounding the battle charge to solve everything. Medical interventions are of course sometimes necessary, but for many patients sincere interest from the doctor is what's more important. Senior citizens often go for days without talking to anyone about their lives or interests. Just imagine how lonesome that must feel.

Listen first, then decide whether any action is necessary: that's what I learned from those Friday mornings. I came to realise that our profession is a very reflective one, and that in a nursing home, medicine should be at the service of promoting a meaningful life.

I visited that man nearly every week for two-and-a-half years. During that time, he opened up and started talking about his wife, eventually even asking things about me – how my week had been, or how my holiday was. When my studies eventually sent me to another ward, I decided to continue our sessions anyway. I didn't want to lose sight of him.

Then, one Friday morning, his bed was empty – he'd died unexpectedly the night before. I was upset, for I had come to value his company, and he mine. Perhaps our talks had made him feel a little special after all. In the nursing home he was just one of many, and his crankiness had made him quite unpopular. Loss of status is common among the elderly, especially when dementia sets in. There is nobody to offer the odd compliment, nobody to affirm their place in the world, to say well done, or that they've done a good job. This man showed me just how much a little genuine attention can achieve. If there's someone who makes you feel special, then life is worth living.'

Euthanasia

Ted van Essen, general practitioner

'I could see straight away through the glass that she wasn't doing well. She looked utterly despondent but agitated at the same time. It was a Sunday afternoon, and I wasn't on duty, but I still ran my practice from home back then, and she called me up. I hadn't seen her in a while: I knew that she suffered from recurring episodes of severe depression and that all efforts to treat her until now had failed. During the last episode she'd become a danger to herself and had been forcibly admitted to hospital. Now, it seemed, she was back at home.

I let her into my hallway, and she got right to the point. I want to die, she said, and you have to help me. I could see straight away that she was serious. I knew her history, knew the anguish her depression caused her. She was still young, in her early thirties, but she really was a pitiful case and had been so for years.

But I wasn't going to solve her problems right there in the hallway. This all happened long ago, long before the Euthanasia Act came in. Assisted suicide was far more complex in those days, especially where psychiatry was concerned. I felt overwhelmed. I hadn't been a GP very long, and I needed time to think. In my head I mapped out the whole sequence of events: I would need to speak to her psychiatrist, as well as her family. I suggested that she come back the next day, so that we could talk everything through properly. She agreed.

Looking back, I regret not giving her more time at the outset – I was too easily convinced that I'd reassured her. I thought she understood that I simply couldn't say yes straight away, plus I thought she was doing better since she was back at home. But I was making assumptions. I should have started the discussion immediately, then I would have realised the seriousness of her condition. I let her go based on my initial impressions and her promise that she would come back.

The next day, she didn't turn up. I didn't stress about it – Mondays are always hectic at the practice, and I had so much on my plate that I was grateful for the extra space. I assumed she was safe with her family and that they would contact me if there were any problems.

On Wednesday morning, during my surgery hours, the police called. She had jumped off a building not far from my practice, and they wanted me to come and identify the body. Her appearance would be too distressing for the family. I can still remember standing there in that tiny little mortuary, leaning over her body. The extra day had been one too many.

Her parents are still angry with me. Not because she died, but because I hadn't agreed to the euthanasia. They had shared in her suffering all those years, only to see her life come to such a horrific end. They refused ever to see me again.

It's been twenty-five years now, but this young woman's story shaped my attitude to euthanasia for the rest of my career. Termination of life might be more intricate when psychiatry is involved, but it can sometimes offer a more humane solution, and I've seen the desperate lengths a person can be driven to if that door is closed off.

Whenever euthanasia once again becomes the topic of public debate, she returns to my thoughts. I opened my door to her that Sunday afternoon, which makes me partly responsible for what happened. She interpreted my hesitation as a rejection. The sad truth is that I genuinely wanted to help her. Things shouldn't have ended the way they did.'

Mania

Veerle Bergink, psychiatrist

'Her remarks could be so cutting and vicious – barbs that were deliberately intended to make people feel rotten. I had just started out as a psychiatrist, and if ever I stumbled over my words, she would cruelly ask if I actually knew what I was doing. If I wore anything to work that I was a little doubtful about, you could be sure she would point it out to me.

It was part of her disorder, I knew that. She was a young and well-educated woman, who had been admitted to our ward for mania. Her behaviour had inexplicably become crazed, uninhibited and full of energy, she was convinced that she needed hardly any sleep, her thoughts were unstoppable and erratic. It was a mood disorder, which in her case also had an unpleasant side effect: she became irritable and easily offended, which made her constantly angry and annoyed. She could be so nasty and cruel, not only to us but also to her husband – a charming fellow who was deeply hurt by her biting comments.

Psychiatrists in training are always taught that a patient's personality cannot be judged during mental states like mania, depression or psychosis – people with psychiatric conditions are not themselves, after all. But with this lady, I had enormous difficulty relying on my professional expertise. I noticed I was harbouring unpleasant thoughts, such as wondering how this truly horrible person could have such a nice husband and a good job. And it didn't help that she single-handedly managed to sabotage my research in the department. She had no desire to participate herself and also set all the other patients against me. I concluded she must have been a nasty piece of work, with an intrinsically disordered personality. I didn't say it out loud, but I certainly thought it.

As psychiatrists, we are accustomed to hearing hurtful things. Patients can be spiteful, and you need to develop a thick skin. I tried to take her catty remarks with a pinch of salt, but I couldn't ignore them completely. Perhaps it was because she looked a little like me . . . Anyhow, she sure knew how to press my buttons.

And then she started getting better. The treatment was effective, and little by little an intelligent and engaging woman emerged. Two years after she was discharged, she came back for a follow-up, and I was reminded yet again of how charming and intelligent she was, with a marvellous sense of humour. That's when I saw just how clearly mental illnesses can blow certain personality traits out of proportion, sometimes bringing out the absolute worst in people. Here was a bright, clever woman, whose mania had transformed her witty and incisive remarks into harsh and brutal ones.

That realisation changed me, both as a psychiatrist and as a person. Mental illness can strike anyone, and it's awful if your true personality is masked as a result. As a psychiatrist I knew the theory, of course, but this patient showed me the importance of putting it into practice and of not passing judgement on someone's character straight away.

In my private life, too, I was always too quick to judge. But I'm a lot less critical now, especially with people I don't know well. One bad-tempered comment or snide remark isn't necessarily the sign of a fundamentally unpleasant character. How people act isn't always an indication of who they really are.'

Young and old

Tommie Niessen, nurse

'It struck me that he wasn't at all judgemental. I look like anything but your average nurse, with my tattoos and shaved hair. But he never said a word about it. I later heard from his daughter that he was a little taken aback when he first saw me, but I never noticed. He asked me what my tattoos meant. I told him that my granddad always used to fold paper boats for me as a kid, and that's why I had one tattooed on my arm. The ink even contained my grandparents' ashes. He listened closely, and was touched. Young people often think the oldies are boring, while the elderly have all sorts of opinions about the younger generation. But it wasn't like that with us. Although we were more than sixty years apart, we hit it off like you wouldn't believe. Sometimes in life you meet someone who you don't need words to understand, and for me that was him. His name was Calum, and he was born in the north of the country, though he had settled in the south for work. Now, because he was ill, he was living with his daughter. His kidneys were failing, but he'd refused any further treatment. I went to see him a few times a week, and we'd have a cuppa and he'd start telling me stories. Man, he was a storyteller – he could talk about life, love and all the rest. Our chats were never just small talk, they were always more like deep-and-meaningfuls. He was very religious, but still open-minded and, like me, interested in spirituality.

A few months later I went on holiday, and then I got a job somewhere else. But I promised to come and see him still. He cried on my last shift. At the end of last year, I got the news that he was in a hospice and not doing too well, and I felt guilty because I'd never gone back to see him. From then on, I visited every week, and we picked up our conversations right where we'd left off.

When his son-in-law told me that he didn't have much longer to live, I went to say my final farewells. By the time I got there, he had already become unresponsive. I sat down by his bed, held his hand and just started talking. Not long after, I saw a tear roll slowly down his cheek. I don't know if he knew that I was there, or whether that tear was even meant for me, but it was such a beautiful moment.

I've met very few elderly people who could talk as openly about their feelings as he could. He was grateful to have lived an exciting life and so happy that there were people willing to take care of him when he got sick. You're all systems go now, he said to me, you're out chasing your dreams, but remember not to be too focused on the destination – the journey there is just as important. As you get older, he said, you'll reflect on your life and realise that you should have stopped to enjoy the view a bit more, instead of racing ahead all the time. He taught me that I should be grateful for what I've got. Appreciate the little things, he said, be conscious of the people around you and of your own behaviour.

I don't always manage to put his advice into practice. I do reflect a lot more now, which is a good thing, though I still have trouble putting the brakes on from time to time. But with him in the back of my mind, it's going better and better. I saved his mourning card, which has a line from the Bible on it: "Without love, we are nothing." And that's exactly why I work in healthcare: because being there for someone else is what love is all about.'

Hope

Henk Eilander, clinical neuropsychologist

'She was underwater for so long in the ditch behind her house, she'd suffered severe brain damage and ended up in a coma. The doctors in the hospital had managed to keep her alive, but that was all they could do. She needed to go into a care facility, but the local paediatric ward was closing down, so she wound up with us in the rehabilitation centre. She was just two years old, unconscious and barely responsive and – most heartbreaking of all – just cried and cried and cried.

We had no idea what to do with her, we'd never had to care for anyone in her state before. This was over thirty years ago: all we could do was wait and hope for a spontaneous recovery, which almost never happened. In the international literature, I read that it was sometimes possible to raise awareness to a higher level, but how? And what to do with the parents, who were racked with guilt and clinging to any hopes of recovery? It was all new to us.

The parents noticed that their daughter was a little more alert at night, and so asked us to do some evening stimulation therapy with her. We tried a controlled multisensory environment (called *Snoezelen* therapy), sang songs and had her touch things, but there was no improvement, and so we eventually stopped. Then the parents learned of a treatment programme run by two American therapists that was said to produce incredible results. Once a month the Americans travelled to Spain to see new patients. I accompanied the parents on the flight there, with their daughter along in a wheelchair. I can still see us now, trudging through that Spanish city, jolting and jittering across the cobblestones. The American treatment proved to be quite intensive: patients needed stimulation for hours at a time, sometimes by five people at once. The therapists said that injured brains could benefit from increased blood flow and advised us to hang the girl upside down for several minutes a day. She was also to make assisted movements, so that the body could experience them and eventually – hopefully – take over. The parents were told there was a chance of improvement, if only they would do their best.

They were willing to try, but we weren't. Some of the American theories were based on no evidence whatsoever, and we feared that they might even be detrimental. We agreed that she would spend half the week with us and the rest at home, where her parents and a group of volunteers would try out the new techniques. I visited them once: they had transformed their entire attic into a private rehabilitation centre.

Several years later, the little girl died. I remember my relief when I heard. All those treatments had never helped a bit, she had become so severely disabled. And she never stopped crying.

Her story had a striking impact nonetheless. That girl and her parents gave us an important revelation: that victims of brain damage must never be denied chances for improvement. That's why we launched our own rehabilitation programme, one offering plenty of ways to stimulate young patients with reduced consciousness, which involves the parents just as much as the child. We take the time to find out what the family's priorities are and to explain the prospects for recovery and what they can do to help.

I know there was nothing that could have helped that little girl, her oxygen deprivation had been so severe. I've never seen a patient make even the slightest recovery from a similar state. But I did learn an important lesson: never take away the parents' hope. The expectations we set must be realistic, but we must never leave them empty-handed.'

Barriers

Hugo van der Wedden, nurse

'She always called me John, and thought I was the boy-next-door she'd had a crush on as a girl of eighteen. If ever she caught me chatting to other women, she got jealous. But apart from that, we got along famously – there was something about her that I liked. The edges of her memory were fraying severely, and in her mind she was a young girl again. It must have been a joyful and lively time in her life. At least, that's the impression I got. We would dance across the ward to the music of Manke Nelis and enjoy the odd glass of advocaat liqueur together.

She had very few visitors. Only her children came by occasionally; the rest of the time she was mostly alone. Shortly after falling and breaking her hip, she died, and my nursing-home colleagues and I attended the funeral. To my surprise, the cathedral was full to bursting. I surveyed the crowd and wondered where they had all been over the last few years. Aside from her own children, nobody in the church had ever been seen at the nursing home.

People gave speeches, recalled fond memories and, to a man, they described her final years in the nursing home as tragic. So the time we shared together – when she was truly happy and we had fun – was to be written off as lamentable and unfortunate? That hurt. Of course she experienced moments of sadness, but usually only because she was so lonely. In the cathedral, that sadness was embodied by all the people who had kept their distance for so long and yet suddenly turned up to pay their final respects.

The sea of faces in that church preyed on my mind. At first, I was angry, but when I later studied sociology, I learned that there was more to the story. The fact that those people had never come to the nursing home was probably related, at least in part, to the culture of the organisation itself. If you want to bring people in, you have to create an environment where the outside world feels at home. The healthcare sector is still highly focused on the individual, on the patient. It's as though we feel powerless to get a resident's social circle involved. For a long time, nursing homes were a kind of impregnable fortress, with fixed visiting hours and families kept at arm's length. Although the doors were flung open decades ago, the barriers surrounding many organisations still seem to be too high. Luckily, there are plenty of ways to change that.

Another of my revelations was that each and every one of those churchgoers must have suffered a little tragedy in their relationship to that woman. At a certain point, all of them – for personal reasons – had made the conscious decision to stay away. Dementia makes people unpredictable, which often makes others hesitant to approach. It's all so confronting that friends and acquaintances don't know how to react, meaning that people with dementia slip further and further into loneliness. And they know it, I'm sure they do – they're aware that their family and friends are slowly but surely drifting away.

Every once in a while, this lady did realise she was in a nursing home. Whenever that happened, she grew sad and we did our best to console her. I suspect that her life slowly began crumbling apart in the years before she came into care. One by one, all those people in the church trickled out of her life, and it must have been very painful for her. And therein lies the great tragedy of dementia, in how we deal with it as a society. People are no longer taken seriously, and they know it. Perhaps that's why the church was so full: it was a form of compensation, and those who had disappeared from her life came to her funeral to atone for a sense of guilt.'

Lulled to sleep

Joost Drenth, gastroenterologist

'I met him when I was still doing my foundation year. He was in his forties and had had a rough life. Born into a dysfunctional family, he was put into foster care, ended up on the fringes of society and eventually became a drug addict. It was then that he contracted hepatitis-C, a viral liver infection, which is how he ended up in my office. There was only one available treatment back then, but it came with serious side effects and in his case proved ineffective. All his hopes were pinned on discovering a cure. I've rarely had a patient who pursued a treatment so doggedly. He conducted his own research, and our appointments became almost technical discussions. Whenever I came back from a conference, he would ask: "And? Any news?"

As time went on, I began to understand the burden he carried with him. So many of his old friends, who had led similar lives, had already died of hepatitis. He had seen how cruel the virus could be, and to his best friend especially, whose liver and brain had been utterly ravaged. That can't happen to me, he said. He was petrified of meeting the same end.

Several years ago, new drugs came onto the market that had successfully contained the virus within a few months. I was so happy to give him the news, but in his case, again, the medication had virtually no effect. After finishing his treatment, the virus returned. When an alternative drug didn't help either, he was crestfallen. Then a third drug was released last year, which targeted the specific strain of the virus that we knew was in his liver. Initially it seemed to work, but afterwards he displayed new symptoms, so I did an ultrasound. I can still remember the image that appeared on the monitor: his liver was entirely lit up, white and gleaming. I was bowled over. A CT scan confirmed the diagnosis. He had liver cancer.

He was admitted to a hospice five weeks ago. I went to see him; he has a magnificent view of the river. I confessed that while I had travelled alongside him all those years as a companion and navigator, I hadn't been paying enough attention. I should have done an ultrasound sooner – I usually do with all my patients, but with him I'd overlooked it for some reason. I now see why. The two of us had made a kind of silent pact, to rid him of that infernal virus for good, a virus that had become as much mine as his. I had been so focused on that one goal, convinced that together we could do it, that my obsession had eclipsed all else.

He asked whether doing an ultrasound earlier might have helped. Perhaps not, I said, since he had never even wanted surgery, let alone a liver transplant. Still, his case humbled me. I've learned that a broad perspective is always necessary, as well as a certain level of professional distance. And I've gained a real respect for diseases, having seen how they can bide their time, like a waiting assassin, before pouncing unexpectedly on their victims.

He once told me of how proud he was to have quit smoking. It was the last of his addictions to go, and it had been a hard road and was supposed to have marked the beginning of a new life for him. But he now faces the very fear that first gripped him all those years ago – soon he'll be dead, just like all his friends. And I couldn't protect him. He now wants to stick it out as long as possible. We keep in touch via WhatsApp: "Thanks for your message, Joost," he wrote recently. "I'm comfortable and happy here, by the river. I think of you sometimes."

Staying in control

Angela Maas, cardiologist

'She was crazy about her cat, had a devoted daughter, nice neighbours, a fine home and was living the life she wanted. When I identified a serious abnormality in her aortic valve, her response was resolute – there was to be no surgery. She was about seventy years old and quite active, the outcomes looked good, but there was simply no changing her mind. Twice a year she came in for a check-up, so I could at least keep an eye on her. I ran an ultrasound of her heart now and then and managed any symptoms with medication. In the beginning she got right down to business, issuing a do-not-resuscitate order and handing over a euthanasia declaration. At every appointment, she checked to make sure I still had both on file.

Things went swimmingly for years, and I almost started to believe that we'd made the right choice. Although the ultrasound showed a severe narrowing of the valve, she remained content. One morning, after cheerfully walking into my office for a check-up, her breathing suddenly became laboured. Gasping for air she sat across from me, foam flecking at the corners of her mouth. If she'd been at home she would have died, but here she was, choking to death before my very eyes. I knew she was against any intervention – Lord, we'd discussed it endless times – but doctors are hardwired to take action. She'd walked in here happily on two legs . . . surely I couldn't let her leave in a coffin?

I decided to put her on a ventilator, in the hope that she might recover from the episode. She was still responsive and gave her consent. Once the ventilator was on, however, it became clear that she would no longer survive without it. The tube in her throat meant she couldn't talk, so she communicated by gesturing and writing down her thoughts. She was as resolute as ever: this was not my intention, she said, this has to stop.

Letting go of her was hard. I knew what her wishes had been, yet had put her on the ventilator anyway. I'd felt overwhelmed and had given in to my doctor's instinct. Against my better judgement, I hoped she might make it through. Had I been wrong to act that way? I expressed my misgivings to the daughter, who offered me some reassurance: her mother held nothing against me, she said.

The woman decided on a day when the ventilator would be switched off. Her final wishes were fulfilled: she wanted to see her cat one more time and to enjoy a glass of wine. And that's how she went, peacefully, with the cat at the foot of her bed and a bottle of wine on her nightstand. A few weeks later I received a touching letter from her daughter. She wrote of how she'd felt for me, how she could see how much I'd struggled with the situation. She said she was happy with the way things had gone, as it had given her a chance to say goodbye to her mother.

Always respect the patient's decision, that's what I learned from this woman. Even if their choice is completely at odds with your own. Patients aren't our property. They don't belong to us, and they certainly don't always agree that we should do everything in our power just because we can. Although it's more accepted nowadays, and doctors and patients now make joint decisions, when I met her fifteen years ago, doctors were still often the ones in control.

She showed me the direction I needed to take, and her memory has stayed with me ever since. She taught me that sometimes, as a doctor, all you can really do is listen.'

Mother courage

Frédéric Amant, gynaecological oncologist

'Monica was over four months pregnant when she was diagnosed with cervical cancer. From that moment on, a battle raged inside her body between life and death. Treating the cancer required the removal of her uterus, which would also mean losing her baby. And this would be for the second time: her first child had been born prematurely several years before and hadn't survived. A hysterectomy now would save her life but leave her childless forever.

She pleaded with me, asking if there was any way at all to save the baby. If it were to affect her own treatment, then so be it – she was prepared to take that risk. The cancer had been diagnosed by chance, spotted by a nurse during a pre-natal check-up; Monica had no symptoms yet. Her pregnancy is what had enabled early detection of the cancer, and she wanted the chance to give something back: to show her gratitude to the child she didn't yet know, but who had already potentially saved her life. I was sitting opposite a young expectant mother with the beginnings of a baby bump, and my heart went out to her.

The medical literature offered precious little information – only a few examples of pregnant cancer patients who had received treatment and successfully given birth. So we knew it was possible, but there were no statistics on the health of the surviving babies or the mothers' outcomes. We had only a few days to decide, and there was so much uncertainty, but we decided to take the chance.

The hysterectomy was postponed, and Monica received chemotherapy to hold the tumour at bay until her child was old enough to be born.

After a thirty-two-week pregnancy, she gave birth to a son. As soon as the baby was out, her uterus was also removed right then and there in the operating theatre. Little Victor proved to be happy and healthy, giving us the courage to start helping other pregnant cancer patients. Whenever one of our patients had a baby, I would visit immediately afterwards. And every time the baby was fine, I left the maternity ward with a sense of relief.

Many doctors thought we were taking unnecessary risks, but still referred all their pregnant patients to us for fear of any problems that might emerge. Their reticence has gradually faded over time, however. We continue to monitor all our mothers and children and have now collected data demonstrating that children in utero are tougher than we thought. During the first few years of their lives, their development is on a par with that of other children. In the meantime, other doctors have come to realise the prospects for both mother and child and only ever contact us now for extra information or advice.

And so it was that one mother, who had the courage to risk everything for the life of her unborn child, impacted the lives of hundreds of other pregnant women. Monica inspired us to launch an international research project, enabling the production of a database that we now use to help other mothers. I can't imagine anything more torturous than having to decide between your own health and the life of a child. Expectant cancer patients once had no choice but to terminate their pregnancy or to give birth far too soon. Thanks to data collected in recent years, those are fates that can now frequently be avoided. We have literally saved children from the jaws of death.

Monica is now officially cancer-free, her son is fifteen and he is doing fine. Every few years we organise a "family reunion" for all the mothers and children who have passed through our ward.

Monica is always there. Last time, I spoke to a young expectant mother who came to the event seeking courage and advice. She hadn't yet started her chemo, but as soon as she saw so many happy children running around, her doubts disappeared.'

The final say

Kors van der Ent, paediatric lung specialist

'The little boy in the incubator had been fighting for his life for days. He had Down's syndrome and had also contracted a severe infection several days after being born. He was on a ventilator, and we gave him heart medication to keep his blood pressure stable, as well as antibiotics to fight the infection. Every day we ramped up the treatment, boosting the technical artillery around that fragile little body, so much so that I wondered whether we were doing the right thing. None of it seemed to make any difference, and his condition got worse and worse with each passing day. On day five, when I ended my shift at eleven o'clock, I went to his incubator specially to say goodbye. This is it, I thought, tomorrow you'll be gone, I'll never see you again. But when I returned the following morning, there he was. His condition had suddenly improved over the course of the night.

A few years later, on a Sunday afternoon, a four-year-old boy was brought into intensive care. A happy, healthy young lad who had fallen into the pond while playing in his grandparents' backyard. He'd been underwater for quite some time, but they managed to resuscitate him in the ambulance on the way to the hospital. His heart was beating again, but he was in a coma. We spent all afternoon on him, put him on the ventilator, gave him cardiac massage and lots of medication. I feared he wouldn't make it, but to my great surprise and relief, he recovered. He opened his eyes and came off the ventilator. That evening, too, I went to stand at his bedside. My shift was over, and I had popped in to say a quick hello. And then, before my very eyes, his heart stopped. We did all we could, but to no avail – he died right there in front of us.

To me, these two boys together are what constitute my "one patient": two children whose stories are still with me thirty years on, because they so clearly represent the extremes that doctors have to deal with. I was still in training when they crossed my path, and they have formed the guideposts in my career ever since. They taught me a lot about my role as a doctor.

Newcomers to the profession often believe they can transform the lives of the sick; that treatments, pills and operations can make a difference and that your actions matter. But eventually the real-isation will come that you're not all-powerful, that you can't just dictate the course of events, that you are helpless sometimes. With that first boy we did all we could, though it all seemed pointless, and he survived. With the second we worked just as hard but saw the opposite outcome. That's the reality that doctors face. We have all this technical whiz-bangery to help patients, and it can certainly get us a long way. But the reality is that sometimes life holds on unexpectedly, while other times it slips through our fingers. It's impossible to know where the chips will fall.

New doctors need to realise that medicine only has limited power over life. These two children have served as my reference points in that regard. Those boys put me in my place, proving to me that God has the final say over life and death and that we should be humble about the potential contribution we can make.'

An infectious laugh

Marie-José van Dreumel, special-needs doctor

'I was his doctor at the day care centre where I'd started working not long before: a boy of fifteen, with multiple severe disabilities. Intellectually he was functioning at the level of a two-year-old, had severe cerebral palsy and suffered epileptic seizures. But he was such a cheerful lad, with a hugely infectious laugh. Hearing it always put a smile on my face.

At a certain point he started losing weight. That's not uncommon among children with severe spasms, because in a way, they are like elite athletes: their muscles are under constant physical strain and they need plenty of sustenance, especially if their bodies are still growing. We increased his caloric intake and he gained a little weight. Things seemed to improve, and I was happy.

Then one day he stopped laughing. He lost his cheerful demeanour, cried more often and showed other signs that something wasn't right. Since he only knew a few words, he couldn't tell us what the matter was. The entire medical team was called in: the behavioural expert, the speech therapist, the physiotherapist, all of us tried to work out what was going on. Was he over or under-stimulated? Had anything changed at home? Was he comfortable in his wheelchair? But nothing seemed out of the ordinary. I examined him physically and still came away empty-handed. But eventually somebody noticed that he was having incredible difficulty swallowing.

Ordinarily we swallow our food without a second thought, but what we don't realise is that it takes more muscles to swallow than it does to walk. His spasms meant that it was taking him four or five attempts to swallow a single mouthful, and each meal was so exhausting that he needed time to recover afterwards. His entire day consisted of eating and resting, meaning he had no time or energy left to do the things he enjoyed. We spoke to his parents and decided to implant a feeding tube into his stomach. It proved to be the right choice: his mood improved, and he was soon back to his cheerful old self.

Six months later, I popped into the day care centre at the end of my shift, when a musical performance was underway. I surveyed the parents in the room and the children on the stage, all of whose special needs were so visible: wheelchairs galore, children having spasms and wearing safety helmets so they didn't injure themselves. And there among them was that one boy, letting loose on his instrument.

Then one of the children came up to me. Doctor Marie-José, he said, surprised, what are you doing here? He looked around, his gaze settling on the stage, then asked: nobody here is sick, are they?

That was the moment when I realised: this is why I do what I do. That child's concept of disease was so completely different: to him, being "sick" meant lying in bed and feeling miserable. But that boy on the stage, whose medical struggles had been ongoing for so long, was perfectly fine. His remark encapsulated so clearly what our job is all about. It's different from being a regular doctor: we can't cure intellectual disabilities, but we can try to relieve the symptoms they cause. Our objective is to work together to give this group of people the best possible life they can live.

It's a long time ago now, I'd only just started out back then, but the memory of that afternoon has never left me. That infectious laugh resounding from the stage once more and that one comment made it all crystal clear to me: this is where my heart lies.'

Fighter

Dick Tibboel, intensive-care paediatrician

'We saw it as soon as she was born: the facial distortions, the shallow eye sockets, the bulging eyes. Kirsten had a syndrome that causes the cranial seams to fuse prematurely, leaving too little space inside the skull and affecting the shape of the face. She also had a severely deformed windpipe, so we needed to insert a tube into her neck below her larynx for her to breathe through.

She had many operations during the first years of her life. There were times when her condition was so dire, when she arrived with flashing lights and sirens, that we wondered if she would even make it. The rings of cartilage surrounding her windpipe hadn't formed properly – a rare condition that had all of us scratching our heads. We consulted experts worldwide, but nobody knew for sure what the best approach was. Every so often we had to replace the tube in her trachea, and whenever we did, we feared the outcome.

Despite the many cosmetic and reconstructive surgeries, she retained a pronounced facial difference, so much so that people gaped and stared wherever she went. Her mother told me she once saw a bus stop in the street, to let the passengers get a proper look. Her daughter could of course feel this judgement of the outside world, the whispering behind her back. How trying it must have been for a little girl growing up, with doubts such as how do I make friends? Should I go to that party? Will I ever have a boyfriend? But I've come to greatly respect her attitude to life. She's a cheerful soul, and the expression on her face says to people: yes, this is what I look like. Is there a problem?

She ultimately became a nurse and would have loved to work with us here in the intensive-care ward. It had become her second home, in a way. But her turbulent medical history means she now carries multi-drug resistant bacteria with her, which we sadly can't permit inside the ward. It was hard saying no – if there's anybody I don't begrudge an opportunity, it's her. She now works in a residential care group with disabled children.

When she was little, she regularly teetered on the brink of death. Countless times we thought, this is it, it's all over. But she always made it, time after time, almost as if she was the one pulling *us* along. But nature is sometimes on our side – not when she was born, of course, but later on. We regularly had doubts about what we were doing to her. She survived all the major operations, but with her appearance she still needed to learn to survive in the hostile outside world. We asked ourselves whether she might hold it against us later. Would she blame us for having saved her? But she eventually put all our fears to rest, as against all odds and despite our doubts, she was tremendously grateful for all we had done to keep her alive. She regularly sent us post-cards, whenever she graduated to her next class at school or when she went on holiday. Her parents were always extremely proud of her and have never hidden her away.

A couple of years ago, we finally removed the tube from her throat for good. For twenty years we had never dared to, for fear of the medical repercussions. But it was fine: the hole has now healed over, and she can breathe normally through her mouth. It showed us that sometimes you just need to wait things out, instead of charging in, all guns blazing.

I've known Kirsten for twenty-three years now, and her story is that of a true fighter. I've never met a patient with such a strong will to live.'

Lifeline

Rien Vermeulen, neurologist

'He was a retired GP, an old-school doctor who had always gone the extra mile for his patients, day and night. He told me he'd often been called away during Christmas dinner – which certainly had its advantages, he added with a twinkle in his eye. I was his specialist for a while, but eventually my services were no longer needed, and I transferred him back to his own GP.

Three months later he was back in my office with new symptoms. He'd been having trouble walking lately, he said, but when I asked for some clarification he kept searching for answers, as though making them up on the spot. A few months later he was back again, with another problem I couldn't pinpoint. When he came back a third time, I knew something was up. I felt the need to be open about my suspicions, so told him – very cautiously – that I thought his symptoms might be imaginary. He leant back, then looked at me and said, "You're absolutely right."

It turned out that he just didn't want to let go of me, because I was a doctor he could trust. His own GP worked like a bureaucrat, he said, working from nine to five and getting his patients out of the door as quickly as possible – a work ethic that was diametrically opposed to his own. Although he didn't necessarily see the medical profession as a "calling", he did believe that to do the job properly, an office mentality was grossly inadequate. He had always known where the specialists were who could be trusted, and they didn't all work in the same hospital. He had only ever referred his own patients to those doctors and would tell them why. Now that he was retired, he was at risk of losing sight of that network and knew that he might urgently need it himself one day. He believed I would always refer him to the right place, and he wanted to hold on to me, like a kind of lifeline.

I told him that he was welcome simply to come and visit me now and again, without the need to invent any symptoms. Just come and talk to me about your old practice, I said.

It became clear just how much I meant to him, one morning when I found an enormous note hanging from my office door at the hospital: *Call the cardiologist – URGENT*. My patient had been admitted with heart arrhythmia, the cardiologist wanted to administer some medication, but he'd steadfastly refused. I'm not taking anything unless Dr Vermeulen has approved of it, he said. The cardiologist protested, saying that I was a neurologist, what did I know about hearts? But my GP was adamant. The cardiologist was fuming, which I could completely understand. I called him straight away and told him that everything was fine.

I've learned that people sometimes really do need to be taken by the hand. Nowadays, patients increasingly see themselves as independent and able to make their own decisions, but is that really the case? If even an ex-GP has trouble, how is the average Joe supposed to make these decisions? There is no reliable data on the best places for patients to go – doctor and hospital rankings are pretty useless. Patients need doctors to guide them along their way, but that won't work unless the relationship is based on trust. This doctor made me realise just how important that trust can be.

For years I saw him every three months, and I always looked forward to our appointments. He would tell me about old times, about how the profession used to be and about his patients – the anecdotes he told were priceless. Our sessions helped him retain the idea that I was always there for him. And to my great happiness, he lived to be a ripe old age.'

Extra time

Wouter van Geffen, lung specialist

'She had gone to the doctor complaining of back pain. But who would ever have suspected the worst in a young, active, non-smoking woman in her early twenties? When she returned complaining of shortness of breath, a scan revealed what the true problem was: she had a rare and aggressive form of lung cancer, which had already metastasised. The disease was ravaging her body. She went blind, and the cancer cells even attacked her vertebrae, causing a lesion on her spinal cord.

By the time she ended up in my office, she had already been through chemotherapy and radiation therapy. A new drug had just been released that targeted her particular type of cancer, and we were going to try it out. Although we knew it couldn't cure her, it could potentially give her more time. But only days after the treatment started, disaster struck again in the form of severe pneumonia. The antibiotics needed time to start working – time that was in short supply. Her airways were so constricted that she was hardly getting any oxygen at all, and we all knew what that meant: her breathing would become progressively shallower, allowing the waste products to build up in her lungs. Unless we did something fast, she would die that very evening.

There was only one way to get her through that night and the following days: to put her on a ventilator, one with a mask that had to be strapped firmly over her nose and mouth. Seriously ill patients have their breathing controlled by the machine, which is extremely uncomfortable, and often they cannot tolerate the mask. There was a chance she might make it, but it was very slim. Sitting beside her bed, I was racked with doubt. If the ventilation didn't work, we would only be needlessly prolonging her suffering. And even if it did work, would the time gained really be quality time? All I could think was: are we really doing the right thing? This girl should be out enjoying herself, at the pub with her friends, leading a fun life, not suffering in a hospital.

I told her things didn't look good. She was gasping for air, could only say one or two words at a time, but still she was resolute. Do it, she said. I was prepared for the worst but will never forget what happened: she tolerated the ventilator, the antibiotics took effect, and the new cancer drugs started working. The tumour cells were driven back, and after a few days some of her vision returned, she could sit up again and move her legs, and she was no longer incontinent. She was discharged and returned home, where she lived for another eighteen months – extra time that she found valuable and fulfilling. And for this woman, who fought so hard for every single day, eighteen months was practically a lifetime.

That moment at her bedside opened my eyes. I had already made up my own mind about her quality of life, but for her, quality had taken on a very different meaning. It's so very easy for us to say: if ever I'm paralysed, or when things start going downhill, then I'll just bow out. But we often draw that line while we're still healthy. I realise now that we can never know just how far we're prepared to go until our worst fears are actually realised. And once they are, I've learned that the line quickly starts shifting. I now know how hard it is as a healthy individual, and therefore as a doctor, to evaluate where quality of life begins and ends. What seems meaningless to us can mean the entire world to a patient.'

Conundrum

Marianne Wigbers, obstetrician

'It was her first child, and initially there were no signs of any complications. She and her husband had moved to our village a short time before: they both had busy jobs and led full and active lives. Two city slickers, busily pursuing their careers . . . I thought I had them pegged, I must admit, and I was taken aback by the decision they would eventually make.

At the first ultrasound, everything seemed fine. But during the twenty-week scan at a nearby hospital, the troubled ultrasound operator called in the gynaecologist, who diagnosed the problem straight away. The unborn foetus had a prolapsed abdomen: a hole in its belly near the navel, where the intestines had protruded and were now suspended in an external membrane. The child also had a heart abnormality. The gynaecologist's immediate suggestion was to terminate the pregnancy. This is serious, he said, and it's still possible to turn back.

Some parents, I know, would prefer not to risk a life with a disabled child and would choose abortion if the ultrasound showed any abnormalities. Society also seems to have become more critical, and these days, anyone who willingly chooses to have a child with Down's syndrome practically needs to defend themselves. But these two young parents gave a resolute, almost passionate response to the gynaecologist's proposal. Termination was out of the question. No discussion, it wasn't going to happen. End of story.

Their baby was born on Liberation Day in the Netherlands. It was a little girl, whom they named Fiona. I couldn't be there, as the delivery took place in the academic hospital some distance away, but because I was in the neighbourhood for a meeting with my professional association, I decided to pop by anyway. The young mother was recovering from her labour, and the baby had gone straight into surgery. The procedure was a lot simpler than the doctors had first imagined. They successfully closed the hole in the abdomen and found that her heart abnormality was less serious than initially supposed. The parents and I talked about it often afterwards: they'd taken such a strong position that I drew strength from it myself. In choosing their child's life, they sent out a powerful message, a response to a conundrum faced by many young parents. Now that pre-natal testing can identify a range of abnormalities in utero and abortion is meeting with less social opposition, the possibility of a "designer world" seems within reach. But if a child is born with a disability, it doesn't mean that their life will be meaningless. To these parents, deciding against the birth of their child was never even an option.

It takes guts to stand up to a gynaecologist who delivers bad news and suggests an abortion. There's no way of knowing what's coming, or even how serious the disability might be. I had assumed that a disabled child would be unwelcome in the lives of these two busy, hardworking people. How wrong I was. It taught me not to judge a book by its cover and that we can never truly predict other people's choices.

I've occasionally considered confronting the gynaecologist about his initial response to the scan. The problems he predicted never even eventuated: Fiona is now a lively, sweet and happy eight-year-old girl. She's missing her abdominal muscles on one side but manages just fine with physiotherapy. Her heart abnormality eventually disappeared by itself.

Just as long as it's healthy . . . how often do expectant parents hear that nowadays? It's a well-meant platitude that's trotted out all too easily, and it's a fairly ludicrous sentiment, when you think about it. These parents showed me just how vast and unconditional the love for a child can be.'

Small victories

Joost Meesters, nurse

'Richard used to be the head of anaesthesia at a nearby hospital. His organisational skills were legendary, and over the course of thirty years he had become indispensable as the chief theatre planner. A few years ago, his co-workers noticed he was starting to become a bit forgetful. When he himself realised he couldn't keep track of things in his head anymore, he decided to start taking notes. His wife later found entire exercise books full of reminders, written long before she even picked up on the first signs of his condition. His forgetfulness had clearly been an issue for far longer than she suspected.

It was some time before he received his diagnosis. Who ever suspects Alzheimer's in a man still in his fifties? He remained an outpatient for some time, but things eventually became unmanageable at home, and then a spot opened up here with us in our younger-onset dementia ward. On the day of his admission we all stood here waiting for him to arrive, the medical team and his entire family. He came inside, and we made a big party of it. He shook everybody's hand, thanked us for coming, you couldn't wipe the smile off his face. When his wife said that he would be living here from now on, she began to cry. He put his arm around her in comfort, blissfully unaware of the cause of her sadness. It was touching to see him like that; I was moved. Alzheimer's disease is often regarded as an endless vale of tears, but just for a moment, he made us believe the opposite.

We quickly discovered that we'd underestimated his condition, he was much farther gone than we thought. He refused to cooperate and often responded angrily to my offers of help – in his eyes I was just an arrogant little upstart, still wet behind the ears. So my co-workers and I decided on a different approach. This man, who once had been the hospital's great administrator, the organiser of staff celebrations, the linchpin of the Christmas committee and the life of every party, needed to be back in a managerial role. Hey Richard, I said one time, while staring at his shoes: tell me, what should we do about that? Then I let him give me instructions on how to remove them. Or I would give him a jar of apple puree that needed opening, and when he managed to pull off the lid, he would beam triumphantly from ear to ear. Shortly after his admission, new washing machines were installed in the laundry. So he marched over there carrying torn-up scraps of newspaper, then stood – paperwork in hand – issuing instructions to the technicians. He was rejuvenated. Whenever he saw us mid-discussion in the corridor, he would occasionally join us, nod thoughtfully and help decide on who should do what.

In our ward, we see so many people who are in the prime of their lives, but no longer understand the world. They often have young children who are slowly losing their father, or a partner somewhere with a whole life ahead of them and suddenly nobody to spend it with. Together, we try to give the tragedy of Alzheimer's a silver lining. Richard showed me that the condition isn't all doom and gloom, that it *is* possible to achieve moments of happiness. Dementia just means celebrating the small victories – a radiant smile above a freshly opened jar of apple puree is priceless.

Recently he was on his way to bed, when he absent-mindedly wandered past his own bedroom door. I called out to him: "Hey, weren't you feeling tired?" "Oh yes," he replied, "well, you know . . . I tend to forget things now and then." Both of us cracked up laughing.'

Zest for life

Ko Schoenmakers, veterinary surgeon

'She'd been born in Romania and suffered a spinal injury as a puppy that had left her severely disabled: she was both paraplegic and incontinent. A Dutch woman had rescued her, lovingly taken her in and was caring for her at home. She put a nappy on her every day and had ordered a special wheelchair to be made – a miniature cart that supported the dog's hind legs, allowing her to run around. The solution worked well for a while, until her two paralysed legs started wearing out. They were being dragged limply across the floor all day and eventually wounds appeared, which then became infected. The little creature felt no pain and started biting into them, doing even more damage – that was the point when something had to be done.

The first vet she consulted told her that the legs could be amputated, but it was a costly procedure. Another option was just to have the dog put down: she was already in a bad way, and her future seemed fairly grim. The owner had come to me for a second opinion.

When I first heard the story, I was sceptical. A creature in a wheelchair, with a nappy on, and we were supposed to patch her up? Was that really in the dog's best interests? Weren't we maybe taking things a bit too far?

And then she poked her tiny head through the doorway: a young, feisty, light-brown terrier with a mischievous grin. She was full of life, tearing into the room in her little cart – her two motionless hind legs dangling in the back seemed not to bother her at all. I looked into her two glittering eyes and instantly, all my doubts were gone. This was an animal that needed our help.

We launched a crowdfunding campaign to try to get her the best possible treatment. Our surgery was free – we charged only for the materials – and we found a prosthetist who was willing to make custom prostheses at cost price. They were necessary in order to protect the leg stumps and to secure the dog's hindquarters firmly to the cart. Some additional modifications were also made to the wheelchair.

It was a huge success. Afterwards she could tear around the room even faster, so fast in fact that she sometimes got snagged – then she would take a few steps in reverse, before racing ahead once more. It was marvellous to see how animals adapt to their circumstances.

It's not uncommon for me to get emotional about animals, but I will never forget the moment when this dog turned to look at me for the first time. The expression in her eyes was so powerful; she had me from the word go. This cheerful little terrier proved to me that in my work, I really can rely on what I see and feel. I need to maintain a critical eye, of course, and I always think rationally about whether a drastic treatment is in the animal's best interests – but I can let intuition be my guide. Just like the animals themselves: they always rely on their instincts.

The terrier is now a regular patient of mine. Two years have passed since she first bolted into my surgery, and she's still fighting fit. Back when I helped her owner make that life-or-death decision, it felt like a heavy responsibility. Now I know that we made the right choice, and it feels like a reward every time I see her playfulness and zest for life.'

Jumping to conclusions

Rob Slappendel, anaesthesiologist

'She was brought into emergency one night at around ten o'clock: unconscious, covered in blood, her face destroyed and hardly a bone left unbroken. The paramedics said it was a suicide attempt, that she had jumped from the eighth floor of an apartment building. Her husband corroborated their story. She was raced to the operating theatre, where a large team had assembled to try to save her life.

I was only in the second year of my anaesthesiology training, but was already allowed independent duty, so I monitored her on the operating table while the surgeons worked constantly on her all night. There was a generalist, a cardiovascular surgeon, a plastic surgeon, an ear, nose and throat doctor, a neurosurgeon, a dental surgeon . . . I've never seen so many doctors come and go. Halfway through the night I started wondering what the point of it all was. Here lay a young woman who had tried to end her own life, and yet here we were, doing our best to patch it all up again. Why were we going to so much effort? A colleague came to replace me at eight o'clock the next morning when my shift ended, but the surgeons' work was far from over.

They were still going at six o'clock that night, when I returned for my next shift. I was a little worried, since being under anaesthetic for that long can put patients at risk. I raised my doubts with my supervisor, who came to take a look and agreed that the limit had been reached: twenty-four hours after she had arrived, they called a halt to the surgery. I took her to intensive care, where I monitored her condition over the next few days. She needed ventilation but had fully recovered from the anaesthetic. After that, she vanished from my thoughts.

Two months later, when I was doing a temporary placement in intensive care, I was reviewing all the patients' medical files and was surprised to encounter my own anaesthesia report. It had been included in that young woman's file, and she was still lying there. She'd had severe complications and survived some major infections, and she was still unconscious and on the ventilator. What a gratuitous waste of time and resources, I thought.

Several months later she began showing gradual signs of improvement, and we decided it would be safe to take her off the ventilator. It was great news, we thought, but after sharing it with her husband, we noticed a marked drop in the frequency of his visits. On the day when the ventilation tube came out of her throat, she quickly regained consciousness. For the first time in months, we could actually talk to her. Her first words shook us all to the core: "My husband pushed me off the balcony," she said.

We were dumbfounded. All those months we thought she'd tried to commit suicide. Police officers came to the hospital so she could lodge an official report and give her testimony. Her husband was arrested and quickly admitted attempted murder.

I felt deeply ashamed of all the negative feelings I'd harboured towards her. From the night of her initial surgery until the ventilation was switched off, I wondered what on earth her treatment had been for. How gravely mistaken I had been. From that point on, my attitude to treating patients changed completely. Whatever a person's age, gender, profession or prior medical history – be it murder, suicide or something else – that was when I realised that everybody deserves our care. This woman's unexpected twist not only changed the way I practise medicine, but also my outlook on life. I now take the time to delve deeper into people's backgrounds and find out what really makes them tick.'

Lack of time

Janny Dekker, general practitioner

'A colleague of mine was on maternity leave, and I had agreed to take on some of her patients while she was away. She would only be gone for four months, so it seemed manageable. But I soon realised I was far too busy, working long days in the practice alongside my research position at the university. In hindsight I can see that I'd overestimated myself; I just thought I should be able to fit the extra work in somehow.

It was during those four months that an elderly woman came in to see me. I knew her fairly well: her husband had died several years earlier, and ever since then an air of sadness had hung about her. She'd noticed blood in her stools and told me she thought it was probably haemorrhoids, which she'd had before. I examined her and her symptoms checked out, so I left it at that. I later realised I thought she looked a little pale, but I pushed the thought away. I was too busy to let it properly register that she looked generally unwell, plus my schedule was already running way overtime. She was probably just eating poorly, I thought – not surprising, if life is not as fun as it once was.

I saw her again after that for a blood pressure check, but she didn't mention her bleeding again. I should have asked her how things were generally, but I was glad of a short appointment so I could keep some time up my sleeve. Later, she told me she'd noticed how stressed out I looked. But she was a staunch northerner, a woman of few words who never wanted to complain.

When she returned a few months later, she was tired and had lost a lot of weight. She turned out to be anaemic, and a colonoscopy revealed the cause: intestinal cancer, which had already metastasised to her liver. There was nothing to be done.

Walking in to give her the diagnosis, I felt like I was going to my own funeral. It had kept me awake the night before, I felt so guilty that I hadn't ordered the tests straight away. Earlier detection might have made no difference – that's how I consoled myself, at least – but there were still signs that I had ignored.

I was expecting her to lay the blame on me. But in fact, she did the opposite: she could see how distressed I was and began trying to cheer me up. Oh doctor, she said, don't blame yourself, there's no need for that. She personally had no real issue with the situation; she was alone and of advancing years and had enjoyed a good life until her husband passed away.

The roles were suddenly reversed: I had come to console her and now I myself was the one being consoled. Her words brought me such relief – the compassion she showed was truly remarkable. But I also realised that the stress I had voluntarily shouldered was what had made me less alert. And so this old lady also taught me an important lesson, that I need to be stricter about my own boundaries. I found a mentor, who helped me realise both that I should learn to say "no" more often and that reducing quantity almost always results in better quality.

Now, whenever I end up with too much on my plate, I think back to that exceptional patient from the north. She was the one who showed me that in life we all need time to look back, to reflect and to contemplate.'

Sense of calm

Jan Lavrijsen, geriatric specialist

'She'd been lying there unconscious in the nursing home for over five years: a woman in her early forties who'd suffered a serious accident followed by complications in hospital. When I started out as a young doctor thirty years ago, patients like her were lying around in back rooms everywhere. They were a forgotten group, obscured from the view of medical science. Theirs was a hopeless lot, one that could sometimes last twenty or thirty years.

Her eyes were open during the day, but we couldn't make contact. She was often restless and suffered frequent and uncontrolled crying fits. Her arms, fingers and toes were bent stiff, despite the physiotherapist's efforts. She was fed via a tube through her nose, which she often coughed up along with gobs of mucus and, occasionally, blood. Whenever that happened, she would start gasping for breath and turn blue. We feared that one day she might choke to death and that her life might come to an inhumane end.

It became a real ordeal putting the tube back in her nose all the time, so we considered the option of inserting a feeding tube directly into her stomach. That would require surgery, however, and we doubted whether that was really in her best interests. Wouldn't we simply be prolonging her condition? We asked her family and GP about it, and their answer was clear: extending her life needlessly is something she would never have wanted.

We consulted everybody – fellow doctors, the ward team, a

pastoral carer, an ethicist and a legal specialist – and ultimately decided against the operation. We also agreed that the next time the tube came out, we wouldn't replace it. I told the family that the last part was my decision, that ongoing medical treatment was pointless.

We didn't know if we were justified in just ceasing treatment like that, in those days there were no known prior cases or legal precedents to fall back on. I remember the moment when the nurse called me to say that the tube had come out again. For months our discussions had been leading up to this very moment, and now I finally had to put my money where my mouth was. I sat beside her bed, alone, and explained to her how I intended to proceed. I needed to know I was prepared to accept the consequences of my decision and see it through. It was a final attempt to make contact. I told her I was convinced that it really was the best option.

Right then, a deep sense of calm fell over her. One week later – six years after the accident that had left her unconscious – she died peacefully. Her family were sad, but relieved for her. The ensuing legal investigation concluded that all due diligence had been observed.

This woman's remarkable story changed the way I look at medicine. That moment beside her bed was when I asked myself the fundamental questions that every doctor should ask. How can I best help her? What more can I do for her in my capacity as a doctor? The answers and their implicit consequences brought about a change in the way I viewed the medical profession. Rather than "Am I justified in stopping treatment?" the question became "As a doctor, should I really continue treatment without any concrete objective and without knowing whether the patient would give their consent?" We have since trained doctors in nursing homes to make these kinds of decisions, conducted research and are now looking into what constitutes the best care for patients with long-term consciousness disorders.

And so it was that this woman, mother of a family, stood at the birth of a new field. She revealed to me the essence of our work, what it truly means to be a doctor. She taught me that every day, we must ask ourselves whether our actions are genuinely worthwhile. Good doctors know not only when they should start treating a patient, but also when and how they should stop. That is what I learned from her long ago. Sometimes, the best decision can be to do nothing at all.'

Rejection

Tineke Westdijk, medical social worker

'The nurse who called me sounded agitated, almost panic-stricken. She had just assisted with a delivery and the baby had Down's syndrome. You have to come now, she said, the mother wants nothing to do with it, she's rejecting the baby. The maternity ward was on the sixth floor of the hospital; I jumped in the lift and had only a few minutes to think. What should I do? I had no idea what to expect.

When I entered the room, the mother was lying with her back to her baby, refusing even to look at it. The father marched up to me, and before even introducing himself, he spat out a phrase that still gives me goosebumps twenty years later. "This bird has to leave the nest," he said.

I turned to the baby; it was a little girl. The nurse was pacing restlessly back and forth. She had already tried to bring the mother around, explaining that babies with Down's syndrome can be dear, sweet children, but none of it had helped.

I first had a calm talk with the mother. Though she was certainly sad, for the most part she was angry. Throughout her pregnancy she had felt something wasn't right, that her child had some kind of abnormality, but the nurse had always dismissed her concerns. There was hardly any pre-natal screening in those days: she was young, her check-ups showed no sign of any problems, so there was no cause for any further investigation.

And now it turned out that she had been right all that time. Why had the nurse refused to listen? We can't deal with this as parents, she said.

The baby had to be medically examined in the paediatric ward, which was several floors below. Fine, the mother said, take it, the further away the better. Her attitude sent shockwaves through the hospital, doctors and nurses were outraged. How is it possible, they thought, for a mother to reject her own child this way? They said we should call the child protection services and have the baby taken into immediate custody. I initially got caught up in the stream of emotions, but very quickly had some doubts. If only we gave the parents some time, I thought, if we just took the pressure off . . . perhaps things might work out.

One day later I accompanied them to the paediatric ward, where the mother saw her daughter for the first time. Three minutes she managed, then had to leave. Little by little, I tried to help strengthen the bond between parents and child. I pointed out the little things, asked them which parent they thought she took after. At first the father was guided by his wife's aversion, but gradually he started developing an attachment. After a few days he took his first photo, which proved to be a real milestone. After that, things steadily went uphill.

Don't be too quick to judge, that's what I learned from these parents. Bit by bit, I slowly found out what the matter was. Their rejection of the baby was motivated not only by anger, but also by fear: they were deeply misinformed regarding Down's syndrome. They were afraid of what kind of life awaited their child and scared that they would be unable to care for her. So we watched some videos of older children with Down's syndrome, and the nightmare they had envisaged slowly faded away.

Since then, I feel more confident trusting in my professional intuition: exactly what I had hoped would happen, eventually did. It took two months, but the parents ultimately picked up their daughter from the hospital and took her home. After a year, it was like nothing had happened: the little girl started laughing and babbling, and they formed a bond. Every child deserves loving parents, but I learned that sometimes it just can't be forced. In her case, love just needed time – and time did its job, in the end.'

Doctor becomes patient

Warner Prevoo, intervention radiologist

'It was only supposed to be a quick scan. For about three months I'd had a weird cough but hadn't paid it any mind. I assumed it was stress-related, as there had been some hassles at work. My GP suspected pneumonia, but antibiotics didn't help. When I started feeling seriously unwell, I decided to bike down to my own hospital for a CT scan. Just in case.

I joined my colleagues in the results room afterwards, where the fragmented images of my lungs appeared on the screen – images that I had seen only too often myself as a doctor. I could see the problems straight away: a bacterial infection in the top that was healing nicely; below it to the left was a large blob, with some dark spots to the right. The diagnosis was clear, I had late-stage lung cancer.

The lung specialist I had so often consulted as a colleague suddenly became my own doctor. I already knew the statistics like the back of my hand: my chances were minimal, with only a few per cent of patients still alive after five years. But further testing revealed that my tumour was in fact treatable, and the cancer responded extremely well to a new drug on the market. I won precious time. The cancer has come back twice now; last summer a part of my lung was removed, and I had radiation therapy a month ago.

For sixteen years I treated cancer patients on a daily basis, and now I'm one myself. Since experiencing what patients have to go through first-hand, I've changed. Only now do I see that doctors have no clue whatsoever how patients are feeling. We tend to hone in on the medical aspects, it's all about combating the disease. To us, for example, it's perfectly normal to insert a hollow needle into a patient regularly and extract tissue from their lungs or liver, to see how the cancer cells are responding to the treatment. I recently had just such a biopsy, performed by a colleague who is one of my best friends. It was horrible: I had no choice but to lie there in complete surrender.

All of our words and actions have an inestimable impact on patients, an aspect that is often overlooked. I think it's partly because we simply don't have the time, but also because we can't afford to entertain all the emotions that patients evoke. The green cloth we spread over them before surgery doesn't just define the operating area – it also blocks out the person lying on the table. We subconsciously push them away, despite the enormous psychological impact that disease has on patients. I now know what an emotionally exhausting blow cancer is. You're constantly on edge, you never get another moment's peace.

I had always been a doctor who could cry and would occasionally stew over patients at home. But even so, I could just as easily shrug things off, offering patients some compassion and a few well-meant words of encouragement. But that's empty talk, I have come to realise, and shouldn't be dished out willy-nilly. An outsider can never truly know how a patient feels. Perhaps we should just put it out there in the open: it should be okay for doctors to show their uncertainties and insecurities, to just admit what an awful business it all is and simply promise to do their very best.

Since my diagnosis, I've continued to work as much as I can. I like my job; it brings me genuine fulfilment. It might also be the typical Dutch Calvinism in me: if you can go to work, you go to work. It's been almost three years now, and, truth be told, I've tried to take a more light-hearted approach to my prognosis.

I was devastated when the cancer came back – I'd always counted myself luckier than that.

We're going to keep you with us a while longer, the lung specialist said. I think it would be nice to stick around for another ten years or so. And yet, when I myself say that . . . it does make me incredibly sad.'

Afraid to die

Anne Speckens, psychiatrist

'I was called to his bedside because he was feeling anxious and distressed. Still young, in his early forties, he had been admitted to intensive care with breathing difficulties caused by severe heart problems. The doctors thought he needed a sedative but didn't want to impede his breathing. That's where I came in. I'd only just completed my final exams and was working as an MD in the hospital's psychiatric ward.

His file was an inch thick, and I had very little time to get a grip on his situation. I told him why I was there, that we wanted to give him something to help him sleep better. Then he admitted quite openly why he was having trouble sleeping: he was afraid of dying. If I fall asleep, he said, I might never wake up. The idea that he might never see his wife and children again was so distressing that he dreaded closing his eyes and drifting away, lest it be for the last time.

I was dumbfounded. Nobody from intensive care had made even the slightest suggestion that he might not make it through the night. I was unsure how to respond; his children were still so young, and I think I found the whole situation too painful and confronting. I prescribed my sedative, then left.

The next morning, I returned to intensive care to check on his condition. I looked for his file, but I couldn't find it in the usual place. So I asked one of the nurses. She was very busy and gave a curt reply. "Oh," she said, "he died during the night." I was floored. There I stood, rooted to the spot, while everyone around me was bustling about, already busy with the next patient who would be taking his place. That made sense, of course – IC beds are a scarce resource – but I just felt so lost and alone. I never even brought it up with my supervisor afterwards. My prescription had been in order, that was all that mattered.

My experience from that night has shaped my career as a doctor. I fulfilled my duty by prescribing him a sedative, but I failed him in one essential respect: he needed somebody at his bedside who was willing to share in his realisation that he might not survive. I'd had no idea what to do for him, I only felt helpless and inexperienced.

As doctors, we generally learn how to help people live, but we hardly ever learn how to help them die. We'd all rather not talk about it, not even in psychiatry. Admitting the possibility of death means opening yourself up to other people's suffering and requires a willingness to entertain the notion that life is finite. A lot has improved since I did my training, but young doctors could still do with a lot more support – so they don't go to pieces at the first patient who faces death, like I did.

That morning thirty years ago was when I realised: this isn't what I want, I don't want to get so caught up in medical technicalities that genuine connections pass me by. Mindfulness has since become my main field, and I'm convinced that those experiences helped guide me towards that choice.

I now clearly realise why that patient affected me so deeply. I had only just started out and was full of insecurities. I felt so deeply alone after he died, an aspect of the job that I believe warrants greater attention. Medicine is an emotionally gruelling profession, and there is still so little support for young doctors. We should be looking out for each other more. Take some time, have a cuppa with a colleague now and then and be brave enough to lay your own fears and vulnerabilities out on the table.'

AIDS

Sven Danner, physician

'The nurse asked me to pop in and see him, so that afternoon I stepped into his room, quite unsuspectingly. He was alone. Several years before he had contracted the mysterious new disease that had come to be known as AIDS. His immune system was in tatters, which had left him with severe infections in his eyes and brain. Now he also had a herpes skin rash and a bowel infection for which only an experimental drug was available. I had arranged for his treatment with the manufacturer, which is why he'd been admitted to my ward.

Once at his bedside, he surprised me with an unexpected question. He had come to trust me greatly, he said, which is why he felt comfortable asking me to help him put an end to his suffering. I can't go on, he said, I'm sick to death of being sick. Perplexed, I replied: but we were going to try that new drug, it might get rid of the bowel infection. He told me to sit back and take a good look at him. I'm almost blind, he said, I've been bedridden for months, my skin is falling apart. I can't live without strong painkillers, and now my diarrhoea is so bad I can't even hold it in – the room stinks to high heaven. I can't bear the thought of visitors anymore. And the worst part is that there's nothing you can do: the illness will get me no matter what. There will always be new problems, new infections, maybe even cancer. I surrender a little piece of my life each day.

Back then, it was virtually unheard of for patients to try to dictate the course of their treatment. Our profession maintained a strict hierarchy: doctors knew what was best for their patients, and that was that. True, we had no way of combating the deadly new disease, but we did our level best to alleviate its effects, one by one. And now, suddenly, here was an intelligent fellow, a journalist, who didn't want what I wanted. And he was right: I hadn't seen the wood for the trees, I'd been tackling each of his problems individually and failed to see the big picture, how hopeless his life had become as a whole. I needed to stop doing his thinking for him, I had no idea what patients like him were going through.

I granted his request. It was the first time I'd ever performed euthanasia, and he died calmly and peacefully. It was in the time before the Euthanasia Act, so there was lots of red tape, but due-diligence regulations were already in place, and I made sure to follow those.

Since then I always strive for meaningful discussions with patients: instead of just dishing out information, I also find out how they feel about their own situation. Whenever I suggest a new treatment, I cover both the advantages and disadvantages. It hasn't always been easy, as doctors are expected to be neutral and comprehensive and always to put the patient's interests first. But doctors also have interests, especially when treatment and research overlap, as they did in the early years of AIDS. We were constantly testing new approaches, trying to involve patients in studies as much as possible and of course publish the results – preferably in class-A journals. It helped that I kept that in mind.

Meanwhile, thirty years have passed, and the condition is now manageable. Thanks to patients' associations and the internet, knowledge of HIV has grown to such an extent that GPs in their surgeries are often less well-informed than their patients. But data isn't everything. I now regard the exchange with this patient – one of the very first AIDS patients I ever treated – as a turning point. It made me realise that the most important thing is not the facts themselves, but what they mean to the patient. That realisation has stayed with me ever since.'

Invisible

Arnold van de Laar, surgeon

'She was my seventh patient. I'm certain, because right before going into theatre she asked me if I'd ever performed the operation before. Sure, I said, six times already. Oh, she said, that's alright then. She was a young woman with morbid obesity, who had come to our hospital for a stomach reduction following many years of fruitless weight-loss attempts. I'd only been performing the procedure for a few months, at the request of some of my colleagues.

I was none too keen, to tell you the truth. When operating on patients with other conditions – like a hernia, for example, or cancer – I could always imagine that their fate might also befall me one day. But not obesity. I've always had a relatively healthy weight, and I found their situation almost impossible to relate to. Once the health benefits of a stomach reduction had become generally accepted – in patients with diabetes or high blood pressure, for example – the physicians at my hospital wanted to start offering the procedure and asked me whether I would consider doing the operations.

I agreed but did have lingering doubts. Was it something I really wanted in the long term? Should I really be operating on patients whose condition I didn't properly understand? Rationally I knew I was contributing to their overall health, but I found it hard to empathise with their specific situation.

Then, eighteen months later, my seventh patient came back to see me. She'd lost sixty kilos and had taken part in the *Dam-tot-Damloop*, a sixteen-kilometre footrace from Amsterdam to Zaandam. I complimented her and asked her how she was doing, and she told me people hardly recognised her anymore. She even regularly got asked whether she was new at work and had to explain that she'd been working there for years.

It had never occurred to her before, but her sudden drop in weight made one thing painfully clear: that people had never noticed her when she was so overweight. They'd never approached her back then, and now they did. She'd been bullied in primary school, which was horrible of course. But far worse, she now realised, were all the classmates who had simply ignored her, as though she didn't exist because she was so fat.

I thought back to my own primary-school days. There had been a girl in my grade four class who got bullied a lot. Although I was certainly never one of the perpetrators, I'd never spoken a word to her either. How lonely she must have felt. There were no anti-bullying programmes in those days, and the teachers did nothing to help. She eventually left the school.

I still remember that appointment like it was yesterday. She was in tears because she had suddenly realised what she'd been missing out on her whole life. And, sitting opposite her, I realised that we all have our prejudices: people who look different are clearly not worth the effort. I'd never truly thought about it until that point.

That conversation was my watershed moment. It's been eleven years now, and since then I've performed over a thousand stomach reductions. That one patient helped me comprehend the kind of silent agony obesity can cause. Her words have since been echoed to me by countless other patients: if you're overweight, people see straight through you. Well, not literally, of course.

I now gladly walk into my operating theatre. Of course all doctors have the urge to do good, but I want to derive my conviction from the patients themselves, not from an abstract theory or a textbook. It's now clear to me just how great the consequences of my work can be. While my patients do become physically healthier, there's more to it than that, something worth potentially even more: it's the chance to make new connections, gain new friends, be noticed. It opens the doors to a more beautiful and fulfilling life.'

Right and wrong

Manon Benders, neonatologist

'The little girl was extremely premature and weighed barely a kilo. Her condition deteriorated by the hour, and soon she was taken to intensive care, where she first suffered a collapsed lung and then a severe cerebral haemorrhage. Her fragile brain was so damaged, we all wondered whether we should continue treatment. We knew she would retain multiple disabilities and foresaw an arduous road ahead; hers was a very unfair start to life. We spoke openly with the parents and expressed our concerns: we recommended stopping the intensive-care treatment, in which case their daughter would probably not survive. The parents were livid. This child is welcome, they said, disabled or not. We asked doctors from another hospital for a second opinion, and they echoed our sentiments. But the parents stood firm. They were gravely offended that we were prepared to give up on their child – so offended, in fact, that they demanded a different attending physician. If that's what you think, they said, then you're unfit to care for our daughter. Under the circumstances, I could certainly understand where they were coming from.

The girl survived, but with serious complications. She was often in surgery and spent long periods in hospital. I eventually lost sight of the parents, until years later when I came across them unexpectedly on the street. I very nearly ran into them. We recognised each other straight away, and the mother visibly tensed up.

In the wheelchair in front of her sat a severely disabled young girl. Her hands, feet and head were strapped in place; she was blind and deaf. I said her name, and the parents were surprised that I still remembered it. I asked how she was doing.

Looking at you makes me angry again, the mother said. She told me they were overjoyed with their daughter, that she responded to the people around her and was always cheerful. Seeing how happy she is, the mother asked me, would you still give us the same advice? I fell silent for a moment. What I saw was a severely disabled little girl, just as we had predicted. But the parents were over the moon with her and took all her limitations on board. If you're all happy together, I said, then you made the right decision. They marched off and left me standing there, speechless. I remember watching them walk away for a long time.

To this day, I regard that encounter and the story behind it as one of the most memorable experiences of my career. As doctors, we often make decisions about life and death, but how do we know when we are right or wrong? If the parents are convinced that they made the right choice, who are we to tell them otherwise?

I now understand just how careful I need to be with my judgements. I had decided that this child's inevitable disabilities would be so burdensome, her life would be a litany of woe, borderline inhumane. But everyone is free to decide for themselves what constitutes a "humane" life. I had started making decisions for these parents before I knew even the first thing about their views. It's a precarious balance, and all too often we see the flip-side of the coin: parents who find they cannot care for a disabled child after all, which frequently ends in divorce. We can't tell parents how difficult their lives will be, nor predict how well they'll be able to cope. But this loving mother and father certainly put me in my place. I now believe that the most important thing is to give parents realistic expectations of their child's future and then to reach a proper and well-considered conclusion together. When deciding on the treatment of a premature baby, I now always ask how the parents feel. I listen sincerely and respect their views – no matter how hard it is, or how much they conflict with my own.'

Protocols

Edwin Goedhart, sports physician

'He'd been injured at a training camp during the winter break. His shoulder was dislocated, and because it was a recurring injury, he needed surgery. The operation went well, but he remained in a lot of pain – so much so that after a few days, on a Saturday afternoon, I returned with him to the hospital. Something just didn't feel right. In A & E, they used an ear thermometer to check his temperature. He had no fever, which meant no blood work was necessary, so they sent us home.

The pain was so bad after three days that he was admitted. His shoulder had become infected and needed flushing out. The next day he wasn't looking too good, so I called the doctor on duty and voiced my concerns to him, but they fell on deaf ears. All the usual checks had been done, and there were no signs of anything untoward. The day after that, he was urgently transferred to intensive care: he was in septic shock, and his kidneys were failing. When his teammates came to visit him, they got the fright of their lives – the infection had blown him up to unrecognisable proportions.

For a month he lay there in hospital. The infection was treatable but had done major damage to the cartilage in his shoulder, and all his movements were extremely painful. Despite long-term rehab, he never played football again in the Netherlands. We searched the globe for a solution, eventually finding some doctors in Italy who were willing to operate again. It was a success, and he made a full recovery.

Nobody was to blame for his misadventures; the doctors had all done their job properly. It doesn't even matter to me whether mistakes were made. What I do think is that we might have spared this young sportsman a heap of suffering if only we had shown more courage as doctors. All the protocols had been followed, all the lists neatly checked off, but the patient himself had slipped between the cracks.

Policies and procedures are indispensable: they give doctors stability and guidance in their work and help clarify the steps that should be taken. But we must take care not to make the rules so restrictive that they stand in the way of proper care. The progression of this sportsman's condition was certainly dramatic, but sometimes an extreme example is necessary to teach us a lesson. In this case, I learned that we should never lose sight of the patient, even if it means venturing into unfamiliar territory. Protocols are based on averages, and their purpose is to help prevent failure, not reach perfection. They deliver average healthcare, not excellent healthcare. Excellence can only be achieved through a sincere concern for the patient.

Deviating from the guidelines requires time and above all else guts. Doctors who do so leave themselves open not only to criticism, but also to legal consequences. Instead of the valuable aids they once were, protocols have now become straitjackets, prescriptive rules from which nobody dares to deviate. Being a doctor sometimes means having the courage to go against the grain.

Looking back, I blame myself for being so easily fobbed off that weekend in A & E. I should have insisted on a blood test. Although the guidelines said otherwise, I knew my patient and could sense that something was wrong.

I still hear from that footballer every now and then. After an extended period of rehab, he eventually returned to his home country, where he saw out his career at a local club. He did it for his children, he told me. Because of his injury, they'd never seen him in action on the field. I once told him I thought he would make a fine coach. He recently sent me a message saying he had become just that.'

Too soon

Jan van den Berg, paramedic

'It was the middle of the night, and a report came in of a pregnant woman with a prolapsed umbilical cord. I assumed she only needed transportation, that our job was to get her to the hospital quick-smart. A prolapsed umbilical cord can be life threatening, as it impedes the baby's blood flow. It's quite normal for Dutch women to give birth at home, and this one was lying on her bed, in the midst of contractions. I looked around, but there was no midwife to be seen. The father-to-be told me he had called the hospital immediately, and they'd arranged the ambulance. I asked his wife how far along she was; twenty-nine weeks, she said. That changed everything; I had to switch gears instantly. The baby was months early.

Thankfully, the baby's head wasn't yet visible. If that were the case, I would have to push it back in, as that's what we're taught to do in order to create space for the umbilical cord. We needed to get her on the stretcher as quickly as possible; the only problem was that she lived on the third floor in a block of flats, and the stretcher was downstairs in the entrance hall. So, step by step, we carefully went down the stairs, pausing frequently to allow for her contractions. On the final stair she stopped and said that she felt something coming. I urged her to keep going, otherwise the baby might be born in the cold air of the stairwell.

In the meantime, my colleague had turned the heating up inside the ambulance, turning it into a massive incubator. Premature babies have thin skin and lose heat very quickly; if the child were to be born on the way to the hospital, it would need to be kept warm. We turned on the siren and the flashing lights and sped towards the hospital. I decided to check for the head again and to my horror I saw not a head emerging, but a tiny foot – it was a breech birth! There, in the sweltering ambulance, I could feel the sweat trickling down my back. A second foot quickly followed. I just hoped that the baby wouldn't get stuck on the way out, but it was so tiny, it was born within the minute.

The baby didn't cry, but instead lay completely motionless on its mother's belly. The mother lifted her head and looked at me and asked, in a worried tone, what was going on. Was it alive? We were only two minutes from the hospital, so I decided to put the baby on a ventilator – very carefully, since the lungs weren't yet fully developed. A medical team was waiting at the hospital, and I accompanied them all into the maternity ward. Suddenly, in the lift on the way up, I saw the baby move – then it let out a healthy cry. I was so relieved.

I've been working in the ambulance service for almost twenty-five years and have never delivered a baby on my own before or since, let alone in an ambulance, let alone a premature breech birth with complications. I sure got lucky, I thought to myself. But later, the paediatrician complimented me on my work, saying that I'd simply done a wonderful job. That was when I realised just what all our training and experience is good for. Paramedics need to perform under extreme pressure, remain alert, think on our feet and sometimes even show creativity. Life doesn't always go according to plan, and through this exceptional birth I understood what it really means to practise our profession: we step into a situation and we act.

In the early morning, I wished the new young mother all the best. I never saw the father again; he had followed behind in his car and didn't arrive at the hospital until later. My partner and I returned to our post, our thoughts reeling. We usually never know what our role in the drama has been, whether we saved a life, or how the story plays out. But one week later I received a birth announcement in the mail, along with a letter from the parents. They wanted to express their thanks – it was a little girl, and she was doing well.'

Hormones

Liesbeth van Rossum, physician/endocrinologist

'For six years she had been battling with symptoms that no doctor, neurologist or psychiatrist could figure out. Physically there was nothing wrong with her; it was her personality that had suddenly changed completely. She'd become agitated and extremely irritable and had developed an odd kind of stutter. Before the referral to our hospital, she'd had twenty-nine servings of electroshock therapy, none of which had helped. Her strange behaviour irritated the doctors and nurses, who occasionally suspected it was all an act, that she was inventing her symptoms. An experienced nurse once snapped at her, saying: If you were my own child, I'd throw my drink in your face.

After an online search, she'd diagnosed herself with Cushing's syndrome, a rare condition caused by overproduction of the stress hormone cortisol by the adrenal gland. It sounded unlikely, since Cushing's usually causes plenty of physical symptoms, none of which she had. But we persisted with the tests regardless, and it turned out she was right: a scan revealed a benign tumour on her pituitary gland, which had been responsible for the excess production of cortisol. It did seem weird that only her brain had been affected and not the rest of her body.

We removed the tumour, and she showed signs of gradual improvement. And then, just like that, her old symptoms returned. So now we were two years on and back to square one. She was at her wits' end, because – unlike last time – now we couldn't identify a cause, not even after an endless battery of tests. Her cortisol levels were normal, so where *was* that odd behaviour coming from? We did our utmost to get to the bottom of it: I spoke to colleagues from all over the world, but her condition was unique, nobody had ever seen anything like it. We maybe could have developed some new tests or tried experimental drugs, but without any clear leads or a concrete starting point, we were stabbing in the dark. And she knew it. She was a smart woman, who knew full well what was happening in her own mind, as well as the effect she had on other people. She couldn't bear it anymore: not again, she said, I've had enough. She submitted a request for euthanasia, which was approved.

I still remember when her mother called me – on a Monday morning, after the weekend handover – to tell me the procedure had been scheduled for the following day. I later learned that on the morning of her death, the whole family had gathered together around the kitchen table and that her mood was bright and cheerful in the bizarre knowledge that by that afternoon, she would no longer be there. To this day, I still don't know what was wrong with her.

I have a theory that for some reason, the cells in her body did not respond to cortisol – which would explain why she didn't have the typical symptoms of Cushing's – but that her brain cells did. Perhaps they were even over-sensitive. Her tale was a strong reminder to me of how powerfully hormones can wreak havoc with the brain. We doctors often focus so exclusively on the body; we should have more regard for the brain as an organ, one that is as susceptible to disease as any other, no different to a heart or a liver. If a simple hormonal imbalance is enough to push a person mentally off the rails, then we should be looking out more. I would have loved to know what she was like before her condition. The photos she showed me were of a radiant young woman, with a good job, plenty of friends and a relationship, but she lost it all to an inexplicable character transformation. Years later, at an international conference, I heard about a similar patient, the second recorded case in the world. After my patient's death, I talked to her parents on several more occasions – they were brought to tears every time. Ten years have passed since then, but she is always at the back of my mind.'

Fear

Sylvia Huizinga, dentist

'There was a tense look about him when I entered the waiting room. He was very brave to have come, I thought, and I told him so. When you're that afraid, it's a big step even to make an appointment, let alone actually turn up. As dentists we're accustomed to anxious patients, and I did my best to make him feel at ease.

I'd graduated not long before and had decided to work abroad for a few years. I did what I'd been taught at university: first you explain what you're going to do, then show them the equipment, then make a gentle start. He hadn't even come for anything complicated, just a check-up, but it was still like pulling teeth. No matter what I put in his mouth, he resisted, especially when I worked on his lower jaw. I didn't want to push him, so I always just stopped before things got out of hand and asked him to come back another time. But he kept putting up such a fight, there was no way I could treat him properly. Eventually he developed a cavity, and a section of his molar broke away. Without proper treatment, I thought, he might start losing teeth. I could tell it wasn't the pain that he was scared of, so I knew it must have been something else.

At his next appointment, he came in with a girlfriend of his. He was visibly trembling. I asked if there was anything he wanted to tell me, and he started a story that eventually made him so emotional that his girlfriend had to take over. It turned out that he had been sexually abused for a long time, which explained why lying powerless in the dentist's chair elicited such a strong physical response. As soon as my mirror touched his tongue or I put a cotton swab in his mouth, his reflexes took over and I had to stop. But he was determined and refused to give in to the stranglehold of his past.

I really felt for him – I could see his pain and distress and wanted to do my best to help him through it. We progressed slowly, step by step. I gave him full control, and if ever things got too much, he would raise his hand and I stopped. It can be quite intimidating, lying in a chair like that with your mouth wide open and a dentist hovering over you. Once he felt confident that I understood his situation, he relaxed a little and could let himself go. He kept his teeth in the end.

I last saw that patient fifteen years ago yet whenever I have a nervous patient, I think back to him. I had only just begun as a dentist and at the time was mainly preoccupied with the nuts and bolts of my job, with solving the oral and dental problems. He made me aware that attention, time and understanding are just as vital to the quality of treatment as the procedures themselves. Say what you'll do and then do what you say: that's the way to win trust. They'd taught us this during my studies, but it only dawned on me fully in practice, when I finally succeeded in giving that one patient a greater feeling of security.

I'll never forget his reaction once the procedure was over: he threw his arms around me, so happy that everything had finally worked out. He came back a few more times for check-ups; after that, my time abroad was up, and I returned to the Netherlands. I never could take away the pain of his memories, but it's still a comforting thought to know that together we managed to loosen somewhat the iron grip of his past.'

Inner voice

Adriaan Groen, tropical doctor

'She came from her village to the hospital by ox and cart. She'd tried to deliver her baby at home, aided by the local midwife, but the birth had stalled. From between her legs hung a limp, white arm and a loop of the umbilical cord. She'd been on the road for ten hours.

Together with my young fellow doctor from the Netherlands, we felt the umbilical cord – the artery had no pulse. We listened for the heart tones using a wooden trumpet, and even asked the maternity nurse to double-check, but none of us heard anything. The Doppler monitor – an instrument used to amplify foetal heartbeats – had a flat battery. The ultrasound machine had been broken for months, though a new one was supposed to be on the way.

A caesarean presented a high risk of infection, as I would need to pull the baby's arm – which was covered in dust, dirt and other nasties – back through the birth canal and out again via the abdominal incision. An infected uterus could result in infertility, or even the death of the mother. I realised there was only one thing to do, a procedure that I always found repulsive to perform and which is too hideous for words. Basically, it boiled down to purposefully mangling the baby so that it can be born vaginally after all. As a doctor in Tanzania, it was a procedure I performed with some regularity, and it always made me sick to my stomach.

Every once in a while, a nurse would return afterwards to say whether it had been a boy or a girl.

Shortly beforehand, we made one last check to see whether there was enough room for our instruments. Then, for some reason, we hesitated – I still don't know why. We changed our minds and, after sterilising the baby's arm and umbilical cord as best we could, decided to try a caesarean after all. A scene then followed that will stay with me until my dying day: the child we eventually took out of her belly gave a brief gasp, then started screaming at the top of its lungs. The baby's wailing was joined in the operating theatre by cries of surprise, joyful laughter and songs of praise to the almighty God. But we, the two doctors, stood there stupefied, in full knowledge of the horrific death that this poor child had just narrowly avoided. Only once the final sutures were in place did our hands stop trembling.

Always listen to your inner voice, that's the lesson I learned. Medicine is full of strict protocols nowadays, and while they've proven their worth, we should never lose sight of our intuition. On that January day in 2003 I stopped performing that gruesome procedure for good, and it's all because of that one mother and her baby.

In the west, disfiguring babies to allow vaginal births has long been ancient history. My experiences have impressed upon me that we should never permit procedures in low-income countries that have become obsolete here. Poverty is no justification for double standards. The medical staff in those countries now also feel the same way – ever since African doctors and nurses discovered that the practice was discontinued here some time ago, it has met with significant resistance.

Many African hospitals now have access to higher quality antibiotics, the infection risk has decreased and preference for a caesarean is almost universal. Disfiguring an unborn child should only ever be considered if the doctors' backs are against the wall and the mother's life is in danger, and even then only if an ultrasound has confirmed without a shadow of a doubt that the child is no longer alive.

The mother spent a week recovering in our hospital, then returned to her village on the back of a bicycle, with her baby in her arms. We never told her of the grisly fate that her child was spared.'

Doubts

Ernst Kuipers, gastroenterologist

'He was a man in his mid-thirties, with young children, who'd been referred from another hospital. His diagnosis was pancreatic cancer, but the question was could it be treated? The only available option was surgery, but since the tumour was so large, operating was impossible. Still, we had some doubts . . . something didn't add up, there was a piece of the puzzle missing. The man was far too young for this type of cancer.

We took a biopsy of his pancreas, ran some tests and made an astonishing discovery: it wasn't cancer at all, but rather a special kind of infection that had triggered an auto-immune response. The condition was rare and precious little was known about it, but what we did know was that it could be treated with strong anti-inflammatories. The treatment took effect, the man's symptoms subsided over the ensuing months and the ultrasounds showed the tumour steadily decreasing in size.

A year later, the man unexpectedly missed his check-up at the outpatients' clinic. We called him up and he said that he'd sought a second opinion from international doctors via his GP. He'd had serious doubts about whether our diagnosis was correct.

The doctor abroad examined him briefly and immediately determined that he had pancreatic cancer after all and needed surgery. Since the tumour had shrunk in the meantime, surgery was now a viable option. I sent a letter to both the man's GP and my international colleague, explaining how and why we had reached a different diagnosis and that it seemed to have been confirmed during the intervening period. If cancer were the problem, the man would probably already be dead by now. I heard nothing for a long time, until eventually I received an extract of a letter from the foreign doctor to the GP. The patient had indeed undergone surgery and suffered severe complications as a result. The pathologist had run tests on the tumour: no signs of cancer, only of a receding infection.

Never for a minute did I think any of this was the patient's fault. On the contrary: I see it as my own failure. The explanations I gave him must have been insufficient, which is why he continued to search for answers. I thought we'd given him incredible news: not only was his tumour not cancerous, it was also benign and treatable. But his diagnosis was so rare that he clearly couldn't find the information he was looking for, which left the door open to doubt. Patients are becoming more autonomous and now look things up themselves, a practice of which I wholeheartedly approve. But it also means the doctor's role has changed. It's not a one-way street anymore: doctors need to sense where their patients are, and look ahead, monitor and educate, to stop them from floating adrift in the sea of available information. Is there anything else you'd like to know? What are your concerns? Do you promise to call me if anything is unclear? These, and many more questions, are essential to ask. I thought I'd covered them with this patient, but apparently not explicitly enough. We were already a year on, and he was doing fine, and yet he'd still somehow felt at a loss. It bowled me over, I have to say.

But I would be remiss if I didn't mention the countless other lessons I've learned from my patients over the years. Hardly a day goes by when I don't learn something new. It's strange, but I remember the silliest things about patients: that one image, that one turn of phrase, that one question . . . sometimes I even remember exactly what bed they were in. All because they touched me in some way. We have such intense contact with patients, during a unique and often highly emotional period in their lives, it gives us food for thought and we start to feel for them, whether we like it or not. As doctors, we are shaped by every experience with our patients.'

A love so strong

Hans Wesenhagen, intensive-care doctor

'Marie had been admitted to the cardiology ward with acute and severe heart failure, where she was immediately put on the waiting list for a heart transplant. Her condition went downhill so rapidly, she needed two support VADs. That's how she came to me in intensive care – a pile of flesh and bones, surrounded by machinery. Complications ensued, and we went urgently in search of a donor heart.

I was in touch with her family daily. The situation was dire, and their grief was palpable. I was honest and open about the fact that she might not make it, while my conversations with Marie herself were brief but intense. She occasionally felt that there was no point to it all, that the donor heart wouldn't arrive in time. I tried to get her to have faith; we seemed to understand each other fairly easily, making few words necessary.

Her new heart arrived just in time. After the transplant, she did her rehabilitation in our hospital and then came back for regular check-ups. She also popped in occasionally just to let us know how she was doing. That wasn't uncommon, plenty of patients do the same.

One year later she threw a party and asked me to come. I was suspicious; was there perhaps an ulterior motive behind the invitation? Did she merely want to show her thanks, or was there more at play? I went along and in the end, I was the last to leave.

We talked long into the night. That's when things truly started between us, but thinking back, my feelings for her must have developed sooner. Not while she was still my patient – I'm certain of that – but in the year afterwards. I realised I'd been pushing my feelings aside. I was head of intensive care, after all, and falling in love with an ex-patient was unacceptable. We went away together a few times after her party and eventually confessed our feelings for one another.

We made a conscious attempt to sift through our emotions, to make sure nothing else was at play: gratitude perhaps, or relief. And although we never did truly figure it out, in the end we decided to go for it and make a full commitment to one another.

After taking the step, I quickly told some of my closer colleagues about our relationship. They were all very happy for me. That's fantastic, they said, good for you. Their responses put me at ease, because I hadn't been able to allay my doubts completely. Marie was no longer my patient, true, but she certainly had been. I was balancing on the precipice of my own ethical boundaries, and I needed affirmation from others that it was all above board.

Where I initially did have some difficulty was with my in-laws, as my presence was a painful reminder to them of a very traumatic period in their lives. For a while I had been the bringer of bad news, and suddenly there I was again, sitting down to dinner as their son-in-law. But not only did we get through that too, they also taught me a valuable lesson. They told me what their experiences in intensive care had been like, how hard it had been for them to make sense of all the information. Through them, I realised that I needed to start listening even more carefully, to explain in even more detail and show even greater understanding. It was love that gave me a unique glimpse behind the scenes, an opportunity that very few doctors get.

When we first met, Marie was thirty-one and I was fifty-nine. I did some calculating and told her that our life expectancies were actually not that far apart. A new heart will generally hold out for about twelve years, by which time I would be seventy-one and probably running out of steam myself. It was a silly bit of maths, but it did offer us some consolation and provided a tentative answer to the question of how much time we might ultimately have together.

We're now far beyond the calculated limit, and the love we share is so deep and strong, we are the envy of our friends and family.'

Taking the plunge

Darbaz Abbas, physician

'I was nine years old when my parents fled Iraq and came to the Netherlands as refugees. I was put into a grade 4 class, dropped in among children who were babbling in an unintelligible language. For the longest time I couldn't participate in class activities. At home, things were very unstable. We lived in asylum-seeker centres for years, and eventually our legal options for residency were exhausted, meaning we could have been sent home any minute. After primary school, I was placed in a preparatory VET training programme. That, the teachers said, was the best I could hope for.

After that I completed VET nursing, then moved on to professional tertiary study. Every step of the way, I had to prove to my teachers that I was up to the challenge, despite my linguistic handicap. It's too much for you, they often said, you'll never make it, just do nursing, that's the best you can hope for. After my tertiary study I took a nursing job in a hospital, but never let go of the dream that many had tried to dissuade me from: becoming a doctor. For me, everyone said, a medical degree was a pointless undertaking. Far too demanding – all it would bring was disappointment.

One day, an old man was admitted to my ward. He was a cranky old sod, spewing opinions and criticism everywhere, but gradually I earned his trust and we developed a rapport. He told me he used to be an anaesthesiologist, and he slowly began to show an interest in me. I told him about my dream and about my background, which seemed to be standing in the way.

He was the first person not to respond with any negativity. Go for it, he said, sure you can do it. He was impressed that I'd come so far already. His daughter was a doctor, and he told her about me, and she too was encouraging. The confidence he gave me has stayed with me ever since. I was certainly a headstrong type and believed in myself, but clearly I still needed the affirmation of an impartial outsider. He gave me the final push I needed. To be eligible for the mature-entry medicine stream, I first needed a more advanced high-school qualification, which I obtained through self-study alongside my nursing work. Then I finally took the plunge and enrolled in medicine. Those were tough years; I had to keep to a strict regimen every day. At the weekends I kept working as a nurse, even during my foundation years, partly because I had to finance my studies myself. Although a general pardon meant that I got my residency permit over ten years ago, I didn't receive my Dutch passport until last year, so I wasn't eligible for partial student financing until the very last minute.

I graduated last year and now work as a physician in the very hospital where I was a nurse for so many years. I work in five different wards, from emergency to trauma to paediatric surgery. It's a perfect learning environment, and I want to do my very best to become a good doctor. Right now, I'm searching for a specialist trainee position, but I'm up against candidates who all have killer CVs and way more work experience than I do. And because of my many detours along the way, I'm also far older than most.

Persevere and never think you're not good enough for your dreams: that's the life lesson I took from that old codger who turned up in my ward six years ago. I never saw him again, and I still wonder how he's doing. I would so love to let him know that I did it in the end, and that I – at least in part – have him to thank for my success.'

Jehovah's Witness

Hans Knape, anaesthesiologist

'They were a lovely married couple with three children, and the wife had fallen pregnant again. She had been referred to me because the placenta was in an awkward position – right near the cervix – which can cause excessive blood loss during labour. We discussed the risks, and I told her that she might need a blood transfusion. Straight away, and as coolly as you please, she told me that this wouldn't be possible. She was a strict Jehovah's Witness, and according to her beliefs, she could not accept blood from outside in any form.

But what if you bleed to death, I asked her, leaving your child to grow up without a mother? I will never forget her answer: I would like nothing more than to raise my four children, she said, but doing it in the knowledge that I've had a blood transfusion is unacceptable. I asked about giving her a transfusion in order to keep her child alive, but that, too, was off the table. Shivers ran down my spine. This young expectant mother was deliberately prepared to take a life-threatening risk.

I raised the matter with my colleagues, which gave rise to some heated discussions. Of the sixty specialists in my department, half were unwilling to go along with her request. A doctor, they said, does not allow a mother of three to die if it can be prevented. It felt as though she was asking us to do our job with one hand tied behind our backs.

But the patient's right to decide is important, and I found that it was something I couldn't violate. As doctors, we also have a duty of care. So after much discussion and debate, I agreed to respect her religious beliefs. She wasn't due for another two months, so we had time to assemble a team of doctors and nurses who were willing to help and be on constant stand-by. The anaesthesiologist was on call day and night, as well as a surgical team and staff in intensive care. At every handover, we asked the question: who is on for the expectant mother?

We consulted the hospital's legal experts, and they felt we'd done all we could to provide the best possible care under the circumstances. If the worst should happen, it would be out of our hands. And yet I wasn't fully at ease; we had everything down in black and white, sure, but what if I was the one who had to stand there and let her die on the operating table, leaving her child without a mother?

I remember the night when they called to say she'd been admitted to give birth. I waited on the edge of my seat, and when we later heard she'd had a little boy . . . well, I've never felt such relief. The birth had gone smoothly, and the child was healthy. The mother lost less than 400 ml of blood, completely within the normal range.

Doctors have an innate urge to do good and treat patients with all the knowledge and resources available. But sometimes we need to accept that that's not what patients want. From this woman, I learned that I need to keep my emotions separate. Of course I was aggravated by her request and the decision she was taking for her unborn child. But I eventually managed to put my aggravation aside, which is precisely what allowed me to help her in the end. At the hospital, we realised we needed to treat her wishes just like any other limitation and modify our treatment accordingly – the "limitation" was just a little unorthodox and not of a medical nature. I went to visit her in the maternity ward afterwards. I briefly mentioned how nerve-wracking the recent months had been for us all, but quickly dropped the subject. She'd just given birth, after all, and she was more in need of a congratulations than a lecture. Looking back, we were glad that we respected her wishes, although of course that's easy to say in hindsight. I'm still astounded by her deep-seated convictions and the consequences she was prepared to accept. It was over ten years ago now, but the memory is with me to this day.'

Out of nowhere

Shoaib Amin, cardiologist

'The parents' email arrived on a beautiful spring morning, with an open-heartedness I had never seen before. I was touched. Two-and-a-half weeks earlier, their daughter of seventeen had died – just like that, out of nowhere. The email gave a long and detailed description of precisely what had happened. Eline enjoyed sport, and on the Saturday in question she'd gone to play hockey. After her own game, she had to umpire the boys' team, not her favourite thing to do. Then she went out to eat and for a trip to the cinema with a girlfriend. When she got home, she relaxed on the couch and chatted to her parents for a while. This girl's final hours slowly unravelled before my very eyes, a girl who I had never known.

She had exams coming up, so her parents said she should probably have a lie-in the next day. When she still wasn't down-stairs by midday on Sunday, her mother went up to her bedroom. Her father and elder brother raced after her when they heard a loud cry; that's where they found Eline, dead in her bed.

Not long after their original email, they were sitting here in my office and telling the same story again, full of emotion but in incredible detail. Like detectives in a mystery novel, they were searching for an explanation, desperately trying to piece together what had happened. Had something gone wrong, or had they themselves done something wrong? They had two main questions: why had their daughter, who was still so young, suddenly died without warning? And was their nineteen-year-old son in any danger?

The autopsy at the local hospital revealed no cause of death; Eline's heart seemed perfectly normal. The cause must have been genetic, and so the parents asked us to conduct an examination. We first looked at all the possible genes associated with sudden sleep death, then those linked to heart arrhythmia and finally, genetic markers for early-onset cardiac muscle conditions. We came up empty-handed. Currently we're in the process of comparing all of Eline's over twenty-thousand genes with those of her parents. They won't give up the search and neither will we.

I'm prepared to do all we can to help the parents, and I sincerely hope we can provide some answers. I understand how important it is for them to find out how this could have happened, despite the potential of exposing a real risk to their son. For some time, they told me, many parents in their village had woken up their children during the night, or had the children sleep alongside them. The impact of Eline's death on the community had been that great.

I often deal with grieving patients but never before had anybody's plight cut so deep. These parents gave such a vivid account, put their emotions into words so clearly, I felt like I was standing right there, in their home, in that very bedroom. Through the power of language, I came to see an existential meaning in their words, a search for the meaning of life.

I still remember going home that evening, after reading their first email, with an irrepressible urge to embrace my wife. My father died shortly before I was born, and the cause was never identified. In a month's time, I will become a father myself. I now realise how important it is to appreciate what you have – in love and in friendships. There's nothing wrong with working hard and striving to reach your goals, but we should start every day with a smile.

That sense of vulnerability was something I had shut out for a long time, perhaps because of all the tragedy I see as a doctor. I kept my emotional distance from patients, as a form of self-preservation. These parents broke through that barrier and taught me an important lesson: the harsh realisation that the things we love are so easily lost.'

A welcome baby

Hieke Aardema, obstetrician

'They already had two healthy children, both girls. Now, for their third pregnancy, they'd had a prenatal scan done to rule out the possibility of Down's syndrome. They did an ultrasound, took a blood sample from the mother and crunched the numbers. There was a one in 1,600 chance – negligible odds, so no further testing was necessary. But one in 1,600 is still a chance, and this child turned out to be just that one. The mother had given birth the week before by caesarean; she was six weeks premature. It was a little boy, born with Down's syndrome and a severe heart abnormality.

That afternoon I went to visit and congratulate the new parents. I didn't know them, hadn't seen them through the pregnancy, but could only imagine how they must be feeling. Nobody does a prenatal scan on a whim: the scans are the product of careful consideration, and by that point many parents have already decided that they would rather not have a child with a disability.

The mother had just arrived home from hospital, but her little boy had stayed behind. He needed to build up strength in preparation for major surgery. When with a heavy heart I warily opened the door to their living room, my jaw dropped to the floor: the atmosphere in the room was not at all what I'd expected. It was a celebration, with streamers and balloons everywhere. The two little girls had made drawings for their brother, and everybody was already besotted with him. I can't remember mentioning the test even once. Pure happiness – that was the feeling that won out.

Two weeks later, tragedy struck: the little boy died during his heart surgery. Words cannot express the parents' devastation. There was a time when they'd thought a child with Down's syndrome would find no place in their lives, but once he was there, once they'd seen him, they fell in love straight away, and he was part of the family.

Their story showed me just how powerful the flipside of prenatal scanning can be. It often forces parents into a choice, the consequences of which they cannot foresee. If, instead of a possibility, their son's condition had been a certainty, these parents might have decided not to continue. Could they ever have imagined how differently they would feel later?

Expectant parents need to realise that prenatal scanning often forces them into impossible choices. What if their unborn child is diagnosed with a disability? What if they can't agree on a way forward? They may face a bitter dilemma, a choice between two extremes, and might never know if they made the right one. It's not uncommon for women to terminate a pregnancy following a scan, only to regret it for the rest of their lives – in every child with Down's syndrome, all they ever see is the child they never knew. Children used to be born just the way they were, and of course it sometimes caused grief or anguish. Technology now enables us to leave as little as possible to chance. But is that always a good thing? It's a question we don't consider often enough. Prenatal scanning is intended to give parents security and certainty, but for some expectant parents, all it brings is new uncertainties.

There could have been no better home for this little boy than with these two parents. Sadly, he never made it there; his nursery stayed empty. And his parents grieved for a son that they had taken straight into their hearts: a little boy with Down's syndrome, who from the very first moment had been so incredibly welcome.'

Service station

Casper van Eijck, oncological surgeon

'She'd suddenly turned a shade of yellow, so they did a scan and now she was sitting opposite me, a timid Moroccan woman in traditional dress. Her tumour was in the tip of the pancreas and easily operable – quite a rarity for pancreatic cancer. I was to discuss the procedure with her: the only catch was that she spoke no Dutch.

Her son had come along to act as interpreter, which meant that she received all the information I gave her second-hand. It produced a delay in communication, which somehow seemed to make my words less weighty. At least, her face showed no sign of emotion; perhaps her son was simply more judicious in his choice of words. Even during that first appointment, I noted how protective he was of his mother.

The operation took around six hours, and I phoned him afterwards to say it had been a success. Shortly thereafter, he was back at his mother's bedside. He helped to feed and take care of her, and he translated everything that was said around her. We were in close contact, and I informed him of every step in both the progression of the disease and her potential recovery. After being discharged, she came back to the outpatients' clinic every three months for a check-up. He was always with her. They never complained if things ran over time, never became irritated if waiting times were long. Their attitude alone showed how genuinely appreciative they were. We live in a society where so much is taken for granted, but these two showed nothing but gratitude for the care we provided. And every time the results were clear, I could see how the boy's eyes lit up.

I needed petrol late one night, so I pulled into a local service station. It was shortly before midnight and raining cats and dogs, so I bolted inside, fumbled for my wallet, looked up – and there he was, sitting there at the till. It was his side job, he said, he needed the money to help look after his mother. I had already been touched by his care and attentiveness at the hospital, but now I was holding back tears. His love for his mother was clearly so great, he'd taken an after-school job in some godforsaken petrol station just to help her.

In Dutch society, Moroccan boys are often seen as trouble-makers, but this young man was the exact opposite. Upon seeing him sitting there, any prejudices I might still have harboured disappeared completely. There are so many people who would do well to follow his example. So often I see older patients come to the hospital alone, because their children are too busy. Sometimes I receive emails from the children late at night with long lists of questions, when they finally do have time. Then they expect me to jump to it, quick-smart.

No, I never asked him where his father was, or if he had any brothers or sisters. I suspect that it was just the two of them living together, but he kept all that to himself. Patients divulge what they want to, and that doesn't always include private matters. If they don't volunteer information, it's inappropriate to push.

I recently told him what a wonderful job I thought he was doing, that I have great respect for the loving care he gives his mother. He humbly accepted my compliment, answering that it seemed completely normal to him. I've lost both of my parents myself, which perhaps explains why I'm so sensitive. Why isn't everybody like him? Surely your parents' health is more important

than anything? I still see them both, that boy and his mother –
only twice a year, because she's doing fine. But I'll be filling up
my car at that service station for the rest of my life, where I saw
him sitting there that night.'

Lack of sleep

Chris Braun, non-practising physician

'He was entirely unremarkable, an elderly gentleman sitting on the bed in his pyjamas. He seemed pretty ragged, as though he'd been lying awake all night. I must have looked much the same: a young GP-to-be, near graduation and exhausted after an entire weekend of looking into children's ears and wiping runny noses.

His wife had called me at around five in the morning; I'd only just closed my eyes after the umpteenth house call and had barely slept the night before. He was apparently experiencing bouts of pain similar to those from the kidney stones he'd had before. The pain was gone now, but still they wanted it checked out. I asked if it could wait a few hours, so that I might get some sleep.

They said yes, so at nine-thirty on Sunday morning I turned up on their doorstep. I mechanically went through the motions of tapping his lower back, feeling his chest, but he said nothing; the pain was gone. Clearly there was nothing untoward, so I recommended he go to his own GP for a urine test, just to be sure. I put my hand on the bedroom door handle, ready to leave, when he spoke the words that sent shivers down my spine: what's this strange throbbing here, doctor? I turned around and saw him pointing at his belly. My doctor's bag fell to the floor, and instantly I knew what the problem was.

I placed my fingers on his belly and felt how they were being pressed centimetres apart, to the rhythm of his heartbeat. It was the unmistakable sign of an aortic aneurysm, a balloon-like swelling in the body's largest artery. Given his symptoms, it appeared to be getting bigger by the minute. When this type of aneurysm bursts, nothing can save you. I immediately called the cardiovascular surgeon at the local hospital. He was in theatre, so I was transferred to a member of the surgical staff. I explained my diagnosis, and through the telephone I heard the surgeon call out: send him on over, it's probably nothing.

That same afternoon, the surgeon called me back and showered me with compliments. The operation had been a success, and the patient was recovering well. Good job, he said, it isn't often you see this diagnosis in someone who is still alive. I myself felt no sense of pride, only shame and relief – a perfectly obvious diagnosis had been handed to me on a silver platter, and I'd overlooked it due to lack of sleep. If the patient hadn't rescued me himself, it would have been the most monumental blunder of my young career. But I could speak of my shame to no one: everybody was impressed that I'd managed to make the diagnosis at all. The other doctors in my practice seemed unwilling to realise that overtiredness was no excuse for me to just lap up some story about kidney stones and not bother to investigate any further.

All of this happened over forty years ago. At that time, we all had weekend shifts from Friday night until Monday morning, and then it was straight back to surgery afterwards. The associated risks were one thing we certainly couldn't complain about, as it was all part of the heroic nature of our profession. But I realised that my future as a doctor would inevitably entail regular periods of little sleep and that I might frequently end up in life-or-death situations. It was a thought I couldn't bear. I had one month to go before graduation, so I finished off my degree, then never worked another day as a doctor. I became an industrial toxicologist instead.

Truth be told, I already knew that being a GP wasn't for me. I could just never bring myself to admit it, until this man gave me the push I needed. For a long time, I kept quiet about how crucial the near miss had been in my decision: people were having a hard enough time understanding my career shift as it was. As a GP I would have ended up washed-out, that much I know. This man protected me from that fate and fundamentally changed the course of my life.'

An unbearable thought

Eduard Verhagen, paediatrician

'She was born with blisters covering her arms, legs and belly, and it was clear straight away that her prospects were bleak. Her skin was so fragile, so brittle, that it fell apart at the slightest touch – the unmistakable symptom of a rare and incurable congenital disease. The skin became easily inflamed, giving rise to life-threatening bacterial infections. Because she had trouble feeding, another hospital had inserted a feeding tube down her throat, which proved to be a poor decision: her mucous membranes were damaged by it, and so her oesophagus was now also covered in blisters.

Bridie was only a few weeks old and was suffering horribly, we could all see it. Our treatments brought more misery than relief. Her dressings were changed every second day, a process that was so gruesomely painful, it took place under anaesthetic. We had no choice but to inform the parents that there was no cure and that their daughter would eventually die – all we could do was prolong her life and try to alleviate her pain. They were extremely distressed. They sent me photos of Bridie in the bath: most babies revel in warm water, but for her it was pure torture. What kind of life is this, they asked, if even a warm bath can offer no comfort?

One day, they delicately posed me a heart-rending question. To them, the thought of subjecting their daughter to a period of extended misery before her inevitable death was unbearable. They

wanted to spare her the torment and asked if I would be willing
to assist with her euthanasia. We brought the whole medical team
together to discuss it. Their request was unprecedented, and we
could see the looming legal obstacles: parents are not allowed to
take unilateral decisions on their child's life or death. We spoke
to the local public prosecutor, who heard us out, but offered no
resolution. I can't give you an answer, he said, I can only get
involved once someone has actually died. Though the parents'
wish was completely reasonable, the potential legal ramifications
were simply too great. We replied that, sadly, there was nothing
we could do.

They left the hospital with their baby and a truckload of band-
ages. Bridie died several months later at home, and it was not a
pleasant death. She needed increasing amounts of morphine to
combat the pain, and eventually she simply stopped breathing.
When we heard of it, we were outraged. We had been unable to
offer Bridie and her parents any help at all; it was an appalling
display of medical practice.

Once again, we arranged to speak to the public prosecutor. In
the meantime, the incumbent had been replaced and his successor
was willing to listen, so he came to our ward to assess the situation.
He informed us of prior instances of doctors who had ended the
lives of chronically ill children and who had reported it to the
justice department. There were twenty-two in total, and in every
instance the department had ruled that due diligence had been
observed: the doctors had done all they could, and death had been
the only way to end the child's suffering. Not a single doctor had
been prosecuted, but the cases hadn't been made public – if only
we had known.

The department believed it was time to introduce some trans-
parency. We were permitted to consult the legal reports, on the
condition that we would publish our research findings in a journal.
We did, and the discussion was opened up both here and abroad.
Four years after Bridie's death, we drew up a national protocol, a

set of guidelines for other doctors who find themselves in the same predicament. We later extended our work to cover palliative care, to help relieve suffering in the final stages of children's lives who have no hope of recovery.

Bridie would have turned eighteen this year. I stayed in touch with her parents for quite some time. To this day they are still proud of her and rightly so. I never could have suspected that one little girl could bring about such sweeping changes. Bridie forced us to consider the lot of children without any prospects for a fulfilling life, and that has changed everything.'

An unforgettable night

Leonie Warringa, trainee physician

'Patrick was young, in his late thirties. His condition was worsening with each passing hour: he had bowel cancer with metastases, and he'd just started a heavy course of chemotherapy. But after the first treatment he'd also contracted severe pneumonia, and because the chemo had effectively nuked his immune system, he was left without any natural defences. We gave him three different types of antibiotics and as much oxygen as we could, but his breathing only became more and more laboured; he was coughing up blood, and his blood pressure was so low that it no longer registered on our instruments. We threw everything we had at him, but nothing seemed to help.

I had the night shift. When I clocked on, my colleague had just told him that the situation was dire, and there was a chance he might not live to see the sun rise. Oh no, he said, not now: I was planning to ask my girlfriend to marry me this week, on the day of our eight-year anniversary. His girlfriend – who was sitting beside him and had suspected nothing – burst into tears. In the meantime, friends and family had started gathering at the hospital to say their final goodbyes. One by one they found out he had just popped the question and that she had said yes. The news spread through the hospital like wildfire. Everybody was moved, and very quickly we all banded together with a single accord: to help him marry the woman he loved that very night.

One of the switchboard operators knew a civil celebrant in a neighbouring town. She was still awake, as her daughter was celebrating after her final exams. She was prepared to come to the hospital to perform the ceremony that night, and at two in the morning she arrived, daughter in tow – they had come straight from the party. There were more than enough witnesses, and the couple's IDs had been fetched from home. And because all brides need something old, something new, something borrowed and something blue, her friends also picked up her new blue high heels and her mother's old wedding ring.

The staff from emergency wanted to quickly decorate the nicest room in the hospital and take them there for the ceremony, but the man was not only shackled to a lot of equipment, he was also too weak to leave his own room anyway. And so, at three in the morning, in a deathly silent hospital, a very special "I do" was uttered in a tiny, overcrowded hospital room: around thirty loved ones had gathered around the bed, with the couple's two daughters aged three and five seated beside him.

That same night, at the celebrant's request, I wrote a formal letter to the public prosecutor's office. The couple hadn't registered their intention to marry in advance – a legal requirement in the Netherlands – so I asked the prosecutor to make an exception due to unforeseen medical circumstances. My colleagues who took over the next morning were glad, and touched, that we'd managed to make such a beautiful dream come true.

Patrick survived the night, and the next morning saw the sun rise with his new bride. His unexpected marriage must have given him strength: one-and-a-half weeks later he went home, with her at his side. He ultimately lived another five months and as a proper husband, since their marriage was eventually declared legal.

This all happened ten years ago, but every time I think of it, the emotions of that night come flooding back. Since then I've realised that doctors are not only there to dole out medical treatments, we can also help patients in the search for meaning in their lives, to help answer existential questions. His family's grief at his eventual passing was alleviated somewhat, thanks to the memory of that unforgettable night.

I recently called up his wife, who answered the phone with his surname. I welled up and could see him lying there, battling bravely and happily on, with his two little girls at the foot of the bed. They may be growing up without him, but now they have a strong mother who bears his name, thanks to the events of that night.'

Courage and conviction

Marcel Levi, physician

'He was a secondary-school teacher, an energetic fellow in his forties. He was practically never ill but had been feeling very tired for a while. He hadn't paid it much heed, but when his nose suddenly started bleeding, he thought he'd better get it checked out. Diagnosis: acute leukaemia. He needed to start chemotherapy right away.

It all went pretty well, I thought. There were no infections, no serious complications, the nausea seemed manageable. However, he found the treatment horrendous, and although he went into remission, the leukaemia returned three months later. His only option was to go straight back onto chemo, but when cancer returns that quickly, the chances of survival drop to below ten per cent. For many patients that's still enough reason to continue treatment, but not for him. His response was resolute: no, he said, not again, it's not worth it. I don't want to live the rest of my life in the shadow of this disease. He was young and still had so much to live for . . . we all wondered if it was the right decision. I was still in training, but the doctors around me were up in arms and even suggested calling in a psychiatrist to see whether he might be clinically depressed and not of sound mind. He asked what would happen if he declined treatment. I said he would die fairly soon but that we could help keep him on his feet as long as possible. I saw him often in the months thereafter. He'd made a list of places he still wanted to visit with his girlfriend, so came by whenever he was planning a trip. I would give him a blood transfusion, along with some stimulants to help keep him active.

He died four months later – four months in which he enjoyed life to the fullest. I spoke to his girlfriend afterwards, and she said that her grief had been eased somewhat by the wonderful time they'd shared together. We had all thought he was crazy to refuse treatment, but maybe it was the other way around. Only later did I realise I'd been looking at him through a purely medical lens. In our eyes, the lack of serious side effects meant his chemotherapy had been a success, but he had an entirely different perspective. As doctors, we often underestimate the impact of treatment on patients: hospital visits twice a week, blood tests, doctor's appointments, then a few okay days before it all starts again. And all the while, the clock is ticking. I wondered: how do the survivors – patients and loved ones – look back on that period in their lives?

So often we read in the papers of how a deceased loved one fought bravely until the very end. But is that really the best option? Doctors are always in "treatment mode"; we're trained to try to heal people. Although we do save lives, for many patients our treatments are hardly a walk in the park. This man confronted me with the harsh reality that treatment is not always the best option. The suggestion didn't come from me – in those days, over twenty-five years ago, the thought of sitting back and doing nothing never entered doctors' minds.

He taught me to be honest and open about the subject. Doctors aren't just here to save lives. We need the courage to tell terminally ill patients that there are alternatives, that forgoing treatment can sometimes improve quality of life. Although doctors are definitely more open to the idea nowadays, I still think that many treatment processes are like a runaway freight train. There's nothing wrong with continuing treatment, but patients should know what they're getting into.

Putting the brakes on in time – that's what it's about. It's just difficult to pinpoint the right moment. Patients often know before we do, they sense and realise much more than we think. This man had the courage of his convictions and was brave enough to make the decision for himself. And that's something I'll never forget.'

A death foretold

Arnold van der Leer, nurse

'He was a middle-aged livestock farmer whose cows had made him sick, so we put him in a quarantined room in the hospital. He had paratyphoid, a serious intestinal infection that had most likely come from contaminated milk. I can still see him lying there in his bed – to this day I know what room he was in.

The antibiotics weren't working, so our efforts were focused on trying to find the right treatment. He was nauseous, with a fever and chronic diarrhoea. He needed a bedpan once an hour, despite the fact that his bowels were already empty. When I started my first night shift that week, he introduced himself and made a curious announcement. He told me that I would be present at his death – which, he firmly asserted, would be in seven days' time.

I didn't worry myself about it. He was in a bad way, sure, but nothing pointed to his condition being fatal. We chatted together, about the farm and his family, but every time our talks would return to his premonition. He counted down the nights, one by one. Standing at his bedside during my last night shift, he added a time to his prediction: by six o'clock the next morning, he would be gone. A haunting sensation crept over me – could he be right after all? I called the physician, who took my report seriously and came over. He examined the patient but found nothing of concern. Call me if you really get worried, he said.

Because he needed a bedpan every hour, I could at least keep an eye on him and monitor his condition. At around three, he said he felt a tightness in his chest. His breathing seemed normal, but I called the physician anyway, who got out of bed and came immediately. He examined the farmer thoroughly, even took an X-ray of his chest and lungs right there, but there was nothing alarming. The patient was given extra oxygen to help him breathe more easily.

At a quarter to six, he called again. I grabbed a bedpan and headed to his room. When I walked in, I saw him lurch diagonally sideways across the cushions, and his eyes rolled back into his head. I dropped the bedpan, sounded the alarm, lowered the bed and started CPR. The physician was there in a flash, along with the resuscitation team. We worked on him for forty-five minutes, did all we could, but he was beyond rescue.

Half an hour later my shift was over, and I went home. It was nearly forty years ago now, but I can still recall the state I was in, how empty I felt. I never spoke to his family, and there was no autopsy. Septicaemia was the presumed cause of death.

How bizarre is it that a patient could predict the very moment of his own death, down to the last minute? That whole week it had been a foregone conclusion to him, which I'd dismissed as incredulous nonsense. I never discussed it with any of my colleagues afterwards. But since that night in 1981, I now know that I should take my patients' suspicions more seriously.

Later, when I started working as a nurse anaesthetist, I regularly saw the same thing during my brief but intense conversations with patients in the moments before an operation. Patients know their own bodies so well, their predictions about recovery and the consequences of the procedure are often right. We have a tendency just to brush them off and try to justify everything rationally.

But we can't control everything, not even with the latest medications and state-of-the-art technologies. I, too, thought we had the upper hand, that we could beat the farmer's infection with antibiotics. Now I realise that there's more to heaven and earth than in our philosophy and that sometimes things happen that we can't explain. And you know what? I find it quite a comforting thought.'

A lone little girl

Hugo Heymans, paediatrician

'She was far from home and all alone in our hospital. Once a week, her parents made the two-hour trip from Drenthe to Amsterdam to visit her. The rest of the time she was by herself, a tiny pre-schooler in her single room – I can still see it now. She had a serious condition: her liver produced no bile, causing severe bouts of jaundice and incredible itching. She lost weight and had a host of other medical complications.

I was studying to qualify as a paediatrician, and she was under my care. I became quite attached to her. I lived close to the hospital, and I made sure to pop in every evening and say good-night. When I was on holiday I had to telephone, otherwise she wouldn't go to sleep. She told everybody that she had a mother, a father, a little brother . . . and Hugo. To her, I'd become part of the family.

Then one day she took a turn for the worse. I remember it like it was yesterday. I stood at her bedside and it suddenly hit me: oh God, I thought, this is it, she's going to die. I bolted through the corridors, barged into my professor's office and burst into tears. He was in the middle of a meeting, and the room was full of men in suits – the supervisory board of the academic hospital was on inspection. Straight away I assumed that my intrusion would cause problems for my professor, but no; he stood up, put an arm around me and walked back with me to the ward. There, by her bed, he agreed with me. I see what you mean, he said, you're right, she's not going to make it. She died that very day.

I went to her funeral, something I've done only rarely in my long career as a paediatrician. Although it was forty years ago, the memories of that day remain etched into my mind. The head of nursing came with me, and together we took the car to Drenthe, to the little village of Westerbork. I had troubling associations with that town, since it was where my parents, brother and sister had been imprisoned during the war before being transported to a concentration camp. We arrived at a local hotel, which had a small adjoining hall with a podium. I still remember how the little girl's father took me aside as soon as I walked in. Come see, he said, there's an open casket. And there she lay, beside the stage, the little girl I had said goodnight to every day for over a year, sometimes even making a special trip to the hospital by bike. I had just become a father to a little girl myself, and I was soon in tears again. I can still picture myself standing at her grave afterwards, beneath a tree. We made sure to find a pretty spot, her father said, as he put his arm around my shoulder. It was almost like he was the one consoling me.

This little girl's story shaped me as a doctor. In those days, being a good doctor meant keeping your emotions at arm's length. But I sincerely believe that if that's what you do, you've missed the point of the profession. Medicine isn't purely transactional; you need empathy to understand what patients are going through.

After the ceremony, I had a long talk to the parents about their daughter, which I hoped would help them bear up. I was one of the few who truly knew what her final year had been like, who could reassure them that she'd felt safe and secure in the hospital and that nothing could have saved her. Later, as a fully-fledged paediatrician and with them in mind, I made it a December tradition to call up the parents of all the children who had died that year. Just to ask how things were and to reinforce the sense that they'd done all they could. I now see how crucial that support is. The death of a child is impossible to process, and as doctors we need to help parents continue living the rest of their lives.'

Peanut-butter sandwich

Meta van der Woude, IC physician

'It was a Tuesday afternoon in the summer holidays when she'd started feeling sick at work and began to hyperventilate. In the ambulance on the way to the hospital, she quickly lost consciousness. A brain scan was taken, which showed no abnormalities, but in the middle of the scan, her heart suddenly stopped. We started resuscitation, then moved her to intensive care, where we kept going. It was hard work – her heart just wouldn't start. I could see where things were going.

Her parents and her boyfriend had rushed over to the hospital, and I asked them to start thinking about organ donation. I can still remember her boyfriend rummaging around in her bag for her donor card. After a very emotional exchange, they eventually gave their consent. An hour after being admitted, the patient died. Before removing her kidneys, we took ten vials of blood so that the nearby academic hospital could determine her tissue type – vital information for deciding on the most suitable recipient.

To determine the cause of her heart failure, we also decided to send blood samples to a toxicologist. The autopsy revealed nothing, so her body was collected by the undertaker to be prepared for burial the following Monday. Then, on Friday afternoon, the toxicologist called us. He'd found traces of caffeine and chocolate in her blood, as well as an unknown substance that couldn't be identified by standard testing. I called the municipal coroner, who came over to the hospital with the police straight away.

The young woman's body was transferred to the forensics institute in the nearby town of Rijswijk. Nothing suspicious was found, but it turned out we hadn't supplied enough blood for a full toxicological screening, and there was none left in the body. Then I suddenly remembered the ten vials of blood we'd sent to the academic hospital for the organ donation tests. We contacted them, and luckily there was enough left over. A few days later the forensic scientists had identified the unknown substance: it was cyanide. The woman must have been poisoned.

Her colleagues were interviewed and told the police that she'd started feeling unwell after eating the sandwich that her boyfriend had made for her at home, with peanut butter and chocolate sprinkles on it. She'd taken only a couple of bites and put it down, saying it tasted funny. Her boyfriend, a chemist, was promptly arrested. The police even went in search of the remains of the sandwich, but strangely they contained no cyanide. So where had the poison come from?

After a lengthy interrogation, her boyfriend finally confessed. He had laced her sandwich with a preservative used in his lab, a chemical that is broken down into cyanide by the body. As a chemist, he knew that he could cover his tracks that way. The remains of the sandwich were tested – and traces of the preservative were found.

The police later told me what his motive had been. He had a personality disorder, and when his girlfriend became eager to get married and have children, he saw no way out other than to kill her. The court sentenced him to time in prison, followed by detention under hospital order. Since then, I've become far more vigilant when it comes to sudden deaths. I call in forensics much sooner and order extra blood samples in case subsequent testing is necessary.

A young woman who slipped through our fingers; our thwarted

attempts to find out what happened to her; her parents' panic and grief . . . it all left my colleagues and me reeling. And to think, that boyfriend at her bedside had looked so gloomy and sad, when all along he'd poisoned her sandwich that very morning. The very idea that we'd stood there chatting to her murderer in the final hours before her death still makes me feel sick. He'd almost committed the perfect crime – but it was the donation of her organs that ultimately unmasked him.'

Doctor and daughter

Irene Koning, trainee gynaecologist

'One-and-a-half years ago, she was taken to A & E for the first time in her life. She'd been out jogging that weekend and on Sunday night was suddenly struck down by excruciating abdominal pain. Scans revealed a tumour blocking her intestines, and during surgery they found out that it couldn't be fully removed. The definitive results arrived three weeks later: ovarian cancer. It was late stage and untreatable.

I'm not her doctor – she's being treated at a different hospital – but I still couldn't do without her in my life as a doctor. That's because she's my mother. And although I do help her with doctor stuff – researching things, following up with my colleagues – our conversations are about so much more than just the medical or scientific side of her condition. We talk about her insecurities and about the future. She tells me she feels betrayed by her own body: that one day she suddenly became a cancer patient, and now she'll be one every day for the rest of her life. I ask her whether she would have done anything differently, now that she knows the end is coming sooner than she thought. She answers resolutely that she's achieved all she wanted to in life, but that every extra day is a welcome gift. And it's those conversations, on topics I wouldn't necessarily raise with my own patients, that I find so worthwhile.

To young doctors like me, a patient is usually no more than that – a patient – and communication is fairly clinical. Not that we lack empathy, but a doctor's main concern is the patient's diagnosis and what can be done about it. We're trained to think that way so that we learn what tests to perform to confirm or exclude diagnoses and to keep track of scientific studies. But my mother has helped me to see my patients through a different lens. They have the same questions she does, they share the same insecurities.

Another thing I've realised is that cancer impacts not only patients, but also their families, spouses, children, colleagues and friends. All of them are equally affected and see their lives transform around them. Just like it's happening to our family now.

I recently went to visit a patient and walked in to find three wailing children at her bedside. The parallel struck me right away: not one week earlier, at another hospital, I had sat like that at my mother's bed with my own two sisters. One day I'm a doctor, the next I'm a daughter. That realisation alone has changed the way I do my job.

The chemotherapy is working. The cancer can't be defeated completely, but she'll get some extra time. She's strong, determined and likes to be in control. But recently she's been experiencing moments of anxiety, since she can feel the reins slowly slipping from her grasp. She asked me to look at her euthanasia declaration – it would make her feel better, she said, for us to read through it together. It was an intense and emotional discussion; like many of the talks I have with my mother these days.

Every conversation with her leaves an indelible mark on me – partly because she is so good with words and can articulate her desires, feelings and experiences so beautifully. If only all my patients could express themselves with such clarity, we could get to the crux of the matter more easily and I would be in a better position to help them.

Because of what is happening to my mother – and to us – I now see that there's an entire life surrounding each and every patient. I'm a better doctor because of it – she's always with me in my office, and I'm there with her.'

Empty-handed

Sander de Hosson, lung specialist

'Scans had revealed a tumour in his lungs, which – thankfully – was operable. A decent operation, and he would be as right as rain. But after the surgeon removed a chunk of his left lung, for some reason we couldn't stop his bleeding. Over the next few days, we were faced with a strange conundrum: here was a man in his early forties, who'd always been in the best of health, but was now uncontrollably coughing up blood that simply refused to clot.

We asked specialists from around the country to examine his case and eventually learned that he was generating antibodies against his own platelets – his own body had turned against him. The tumour had produced substances that had confounded his immune system and triggered an attack. The condition was rare and life-threatening. We gave him a high dose of medication against the antibodies, but it had no effect. Every day we put him on a fresh IV of blood platelets, a lifeline that was supposed to help his blood to clot. But nothing worked. He had spontaneous bleeds, bruises everywhere and was constantly coughing up blood. The drain coming out of his lungs was a deep shade of red.

I was the ward doctor, had just started out, and I saw him every day on my rounds. His wife sat at his bedside, along with his four-year-old daughter, who was happily playing the whole time. He had pinned all his hopes on me and became more desperate with each passing day. He'd clutch my hand, begging me over and over: Please, let me survive, you have to save me, my child deserves a father to see her grow up.

His condition had only been reported a couple of times in the international journals, and once we'd consulted all the available professors – even tried experimental drugs – there was nothing more we could do. That was when it hit me: he's not going to make it. I sat beside him and heard his pleas but could only tell him that the end was inevitable. I was totally honest with him, but man, it was hard.

I do try to keep an emotional distance from my patients – once I'm at home, they're off my mind – but I couldn't let this man go. He was living proof that a death sentence can befall any one of us, from one moment to the next. The fact that he was so young made his case all the more frightening. His sadness and desperation are as vivid to me now as they were eleven years ago, and I can still picture his daughter clambering up onto the bed for a cuddle. How do you tell a child their father is going to die? Where do you find the words, when there is nothing more you can do?

He ultimately accepted his approaching death. He had no choice. Every day I dreaded going to see him – what was I supposed to say? Yet I knew I had to sit there beside him and keep listening, until the final day came. That's the lesson I learned from him, one which has served me well ever since: our profession isn't all just medical and technical, it's also about giving time and sincere attention.

I was with him when he died. He was about to bleed to death, and to prevent him being conscious when it happened, we gave him a sedative to send him off to sleep. After he died, all the doctors and nurses came together to discuss what had happened. We were in shock, and it helped to share our thoughts with one another.

What do you do when you're all out of answers? That's what this patient taught me: be honest, listen and, above all else, never walk away.'

One fatal evening

Bert Keizer, geriatric specialist

'He was from Suriname, in South America, and had come to the Netherlands as a young man. Before he knew it, he was addicted to heroin and the classic tale of misery followed: his life fell apart, he developed a lung abscess and eventually contracted HIV and AIDS. That's how he arrived at our facility, where all the city's hopeless cases wash up. Heroin-addicted sex workers, incurable alcoholics, homeless junkies who can't cope with the harsh street life anymore – those are the ones who wind up on our doorstep.

Street-dwellers have a reputation for being fairly rough around the edges. But not this man; surprisingly gentle and kind, and a good-looking fellow besides, I found him endearing. After all those years on the streets, fighting and scoring drugs, here with us he finally found a place where he could relax and enjoy life. Occasionally he would cook a delicious Surinamese meal for everybody in the department. But the high-octane street life, however destructive, remained a temptation: the excitement, the colourful characters . . . it was exhilarating, in a way. And our care centre was *fucking boring*, as he so eloquently put it. So occasionally he would head back into town, get back on the drugs and fall sick again. When he inevitably returned, we always had a room free for him.

For years it went on like this, until one day he seriously took stock of his life. He realised that all the major milestones had passed him by; he had no wife, no children and no career, only a brother who visited him occasionally. He didn't want to live on the streets anymore but was equally horrified by the prospect of spending the rest of his life in a care facility as a fifty-year-old. So he made a decision: I want to die, he said. He stopped taking his HIV medication, but that proved to be fairly ineffective as a means of expediting death – it would take far too long.

One fatal evening, the head of the night shift noticed his bedroom door standing open, then felt a gust of wind blow down the corridor. His room was empty, and they found him lying crumpled among the bushes on the ground below. He'd jumped down from the second floor and was in a horrible state: unconscious, with a collapsed lung, broken bones and ruined kidneys.

I went to see him in hospital, where he'd been put on a ventilator in intensive care. My heart went out to him. How would he ever get out of this? I asked the hospital staff for a meeting and was summoned before a large group of doctors and an ethicist. His request for euthanasia had never been documented, so I steeled myself for battle. You people might have his body, I said, but I have his soul. If you want to bring him back, he needs something good to come back *to*. Then, before all of my colleagues and fellows, I bared his soul to them, explaining how the only future that awaited him was one of misery and unhappiness. His brother was beside me, and he echoed my sentiments.

They listened, and in the end it was a productive dialogue. His injuries were so extensive, the doctors said, that he would never fully recover. They decided to pull the plug that very afternoon. At six-thirty that night, I got the phone call that he had died. I suddenly burst into tears, which came as a surprise to me – I don't cry that often.

Could I have offered him a different fate? It was a question that I pondered for a long time afterwards. I'd never wanted him to die – which was strangely judgemental of me – but I'd developed a soft spot for him, and it felt like such a waste. He'd never pressed me about it, never demanded my help. That's why he'd had to do it himself, alone, in the middle of the night. The idea still haunts me sometimes.

I thought about his life, just as he himself had once done: an immigrant's life, gone to seed. He was such a nice person, educated and with a fine character . . . where did things go wrong? His brother came by to fetch his belongings, an entire life in a rubbish bag. I suddenly saw so clearly that my own happiness wasn't earned by merit alone, that my life had been blessed by good fortune. And I was deeply saddened by his death.'

Bravery

Dame Sally Davies, haematologist

'She'd already been through so much when I met her for the first time: a young girl of ten, who regularly came to hospital with attacks of pain so severe they could only be lived through with opiates. Even before she started school, Laurel had been diagnosed with a hereditary blood disorder called sickle-cell disease, a condition that would overshadow the rest of her life.

I watched her grow up. She developed from a teenager into a young woman, and my admiration for her only increased as the years went by. She finished school, graduated, started going out, fell in love, all the while checking in and out of hospital. How did she do it? Her girlfriends all went to discos, wore skimpy clothes and drank alcohol, while she was bound by strict rules: stay out of the cold, drink lots of water – anything to avoid one of her dreaded pain crises. I remember once she left the house without a scarf, and then spent a long time waiting at a bus stop out in the cold. After that, she was hospitalised with searing pain in her jaws.

As she gained experience, she gradually learned to cope with her disease. I was there in the background as a kind of tutor-partner. I was tough on her, because I cared. There was something about Laurel that spoke to me. Although her entire life was dominated by her condition (one that she shared with her mother and brother), she always remained so admirably calm. She needed frequent injections – to administer fluids and painkillers or take blood samples – so her veins were all messed up. But she never lost her composure, never complained when we missed the vein for the umpteenth time.

I often sat chatting with her on the edge of her bed. We talked about life, but also about death. I was widowed very young; she had already lost her brother and a nephew. Every year in the hospital, she saw people dying of the very disease she was attempting to defy – people she knew and who were often good friends. In Britain we tend to avoid talking about death, it's kind of hidden. Between the two of us, though, there were no barriers to the subject.

For a long time, I never really understood the severity and extent of the pain she had to endure. After having my first child, I talked to her about the pain of the labour I'd experienced. It was dreadful, I said, and I asked her whether it was comparable to the sickle pain she suffered on a regular basis. She had become a mother herself in the meantime and answered quite casually: oh no, sickle-cell pain is far worse. As a teenager Laurel produced amazing art, in an attempt to convey just what these agonising episodes can feel like. There is a painting of hers that I use often when giving lectures to young doctors.

I became an expert in sickle-cell disease, with one of the biggest clinics in Europe. But my most important lessons came from Laurel: she showed me what it's like to live with a disease that will be with you forever, one that dictates your entire existence. But even when her pain must have been truly unbearable, still she soldiered on. She taught me what bravery truly is.

I once took her with me to a lecture for medical students, where she talked about the impact of her illness and the pain in her life. I asked her, do you think there should be a pre-natal test for sickle-cell disease, so that expectant mothers have the option of terminating the pregnancy? Her response was unequivocal: "Yes," she said. "No child should ever have to go through what I went through."

And yet, in spite of everything, she was determined to make the most of her life. She made it through college, forged a career as a graphic artist, wrote a children's book, found a partner and raised a child. And always with her stoic optimism, without a trace of bitterness.

I eventually remarried and pursued my career elsewhere, leaving my job at the hospital. But Laurel and I never lost sight of one another. She was there when I stepped down as the UK's Chief Medical Officer, and gave an incredibly moving speech about her disease and the role I had played in her life. I became her doctor forty years ago, and to this day I remain proud of all that she has achieved.'

Trapped

Mary Reilly, neurologist

'Christopher's decline began when he was a child. First he had trouble walking, then lost partial control of his arms. When I met him, he had already been in a wheelchair for several years. He most probably had a genetic condition, which is why he'd been sent to me. One thing I noticed was that he always wore cheeky T-shirts with provocative slogans and images – only later did we figure out that was his way of trying to tell us something.

I started extensive genetic testing, and shared my data with international doctors: like a detective, I set out to pinpoint the origin of his condition. To tell you the truth, I enjoyed the challenge. There's nothing more satisfying than being sent a patient because no other doctor can figure out what's wrong. But I never really appreciated how much his disease was affecting him, until one of the nurses in my team who had got to know him told us he was depressed – those T-shirt slogans turned out to be his way of reaching out to us. We contacted a psychiatrist and tried out different medications, but it very quickly became clear that his depression was a result of the life in which he had, in his own words, become trapped. There he was, a lonely guy in his late twenties, slowly growing weaker, and feeling immensely frustrated. It was his greatest wish to complete an art degree, but he was rooted to the spot. To be admitted he would need to complete a bridging programme, which was impossible, confined to the house as he was. There was only one thing that could break his isolation: he needed a car that he could drive himself, even with his physical limitations. But he had absolutely no idea how to go about it. One of the nurses saw his frustration, and refused to let it go. She started writing letters, reams of them, and eventually succeeded in getting him the car. It was hand-controlled and voice-activated, the showpiece of the car manufacturer. He immediately started his art course, was admitted to university, got first-class honours and went on to complete a master's. Last year he launched his own start-up, and is now an established artist: he holds art shows and has even been invited to exhibit in the United States.

His work is absolutely astonishing. Despite being unable to use his hands, he creates beautiful art by gripping the pen between his arms. And his talent remained hidden all those years, trapped behind the barriers thrown up by his illness. It's an incredible story, and a frightening prospect that it might never have happened. How many other patients are out there who, just like Christopher, have so much to give but never get the chance?

Now, thirteen years later, he's nothing like the man he was when we first met. I see him every six months. Very slowly his condition is worsening, he needs ventilation at night, and is softly spoken because his vocal cords are affected. But from the first moment he went to university, his depression vanished. In the beginning he would still draw pictures of himself trapped behind bars, to express how he had felt all those years. But now the bars are gone, and his life has changed completely.

I still haven't been able to isolate the genetic mutation that's causing his condition. But now I realise that there's so much more to patients' lives than just the medical side. Nothing that us doctors ever did for this man revolutionised his life as much as that car did, and that fact really opened my mind. Sometimes patients need something other than just the right drugs or a successful diagnosis to turn their lives around. For Christopher, a car proved to be the best treatment.

I am humbled by his talent. Christopher showed me that while my detective work is certainly valuable, keeping a focus on the person behind the diagnosis is just as important.'

A touching letter

Anthony Fauci, immunologist

'He came to our hospital on a Friday afternoon, straight from the quarantined air ambulance that had picked him up in Sierra Leone. A young doctor, he had travelled to Africa as a volunteer to help with the major Ebola outbreak in 2014. He'd been stationed in the north of the country, at a specialised clinic set up in the critical district of Port Loko. One day, he became light-headed and collapsed. Blood tests confirmed the worst: through his contact with Ebola patients, he himself had become infected. He needed to be evacuated as soon as possible.

He could still walk and talk when he arrived, but his condition rapidly deteriorated before our very eyes. He suffered multiple organ failure and needed life support – slowly but surely, death was approaching. The doctors and nurses brought me daily reports on his condition, but for some reason the arrangement didn't sit right with me. I felt like I should be doing more than just listening to their assessments: I couldn't justify asking my staff to put themselves in danger, providing round-the-clock care for an extremely infectious patient, without doing the same thing myself. So I cleared out part of my schedule to allow me to join the medical team.

Every day for two weeks I hoisted myself into one of those hazmat suits – helmet, goggles, the works – and stepped into the isolation room, like an astronaut during the moon landing. Each daily shift lasted two hours. Those suits are so exhausting, you just can't go any longer, and the risk is too great that you'll make mistakes and expose yourself to danger. For nearly two weeks we cared for him, my colleagues and I, under the most stressful conditions. He was one of the sickest patients I've ever had under my charge. This all happened over four years ago, when there were no drugs to fight Ebola, so all we could do was combat the symptoms. And we succeeded: four weeks later he had made a complete recovery and could go back home to his parents.

All that time he had seen only my eyes through the little glass window in my helmet. Once he was on the mend, we started having chats in his little room, but still I remained anonymous. It wasn't until he got back home that he found out who I was and wrote me a touching letter that I have kept ever since.

In the letter he confessed that he looked forward to my arrival every day, that my smile behind the mask had given him strength and that he had enjoyed our conversations very much. Now that he knew who the man behind the mask had been, he felt ashamed at the casual tone he had struck and thought perhaps he should have been more respectful or formal. He thanked the team, saying that without us he just knew he wouldn't be alive. And yet his conscience bothered him: why had he been the one to receive such excellent treatment, a privilege that is still denied to so many African patients? He hoped that at the very least his time in isolation had taught us something about Ebola that would enable us to improve care for other patients in the future.

His case did indeed improve our knowledge of the disease. We had always thought, for example, that organ failure was the result of dehydration and the drop in blood pressure caused by vomiting and diarrhoea. But even though we were able to keep his blood pressure up, his kidneys still failed, along with his lungs, heart and nervous system. His case helped us realise that the Ebola virus is simply an extremely destructive pathogen, mercilessly ravaging everything in its path.

Although he has physically recovered, his experience has left him with a form of post-traumatic stress. He's well aware of how close to death he came and of how miraculous his recovery was from one of the deadliest infectious diseases. He is an illustration, once again, of the almost boundless limits of human resilience: he survived a harrowing ordeal and yet is still able to reflect on and appreciate the efforts of those who helped him through it.

He described a photograph that was taken while he was still on life support in his room, with me standing beside him in my moon suit, and he said he had come to treasure it. He cited Hippocrates: "It is far more important to know what sort of person has the disease, rather than what sort of disease the person has." You treated me like a person, not a disease, he said. And that's what medicine is all about.'

My darkest hours

José Schroe, intensive-care nurse

'She'd already been a patient in our hospital for a few days, and when she was transferred to intensive care that Saturday morning, she knew what the possible outcome could be. She'd messaged me before my evening shift, and when I saw her that night she had a raging fever, shortness of breath and was suffering quite a bit. Meta, the intensive-care doctor I had known for over twenty years, was now suddenly a patient in her own ward; she had fallen victim to the new virus, whose devastating effects she knew only too well.

She could sense that she would need to be ventilated at some point, and was terrified of what might happen. There was a chance she would never wake up from the sedation, but her husband, children, parents and other loved ones weren't allowed to see her because of the infection risk. She was all alone, and it was a heart-wrenching sight. She became very emotional and started crying. Which pain was greater, I wondered, the fear of approaching death or the anguish of not being able to say what you want to say to your nearest and dearest?

"Why not record a few video messages on your phone?" I suggested, thinking it might calm her down a little. I asked if she wanted me to leave, but she said no, I should stay. So I sat at her bedside, cocooned in protective clothing, and held her phone steady while she recorded the video messages. Her overemotional state from before was replaced with complete calm during the recordings.

It was as though she wanted to protect her loved ones, to avoid making things more difficult for them. I felt like an intruder; I didn't want to listen to her confessing her innermost feelings, but I had to. And so it was that I became a party to the most intimate moments of her life. I didn't understand how she could remain so composed when she didn't know if she would ever see her family again. I had a stern word with myself, saying I had to keep it together, it was my job to be there for her. But I remember swallowing and swallowing just to keep myself from choking up. I have hidden away those moments at her bedside in some deep pocket of my brain. Because if I think of it, I choke up all over again.

Things progressed quickly after that. Before being sedated, she left us with some instructions: she told us which catheter to use and reminded us to feel whether she had cold legs. Retaining control until the very end – amidst all the tragedy, it did give us a reason to smile. Soon she was lying there on her belly, silent, asleep and motionless, and calm descended. We had done all we could, now the rest was up to her.

My colleagues found it emotionally gruelling to treat her, and so that very night she was transferred to another hospital. I initially didn't understand why – she was one of us, surely we would take good care of her? But after thinking it over for a little while, I came to realise that they were probably right.

During that shift, I noticed for the first time that I had to shut out certain feelings and observations now and again in order to keep going. Never before had I had such a hard time caring for patients as during the corona crisis. That image of Meta, lying there all alone and fighting for her life, really hit home. The doctors and nurses are praised as heroes for the work they do, but the real heroes were the family members sitting at home, scared and worried, separated from their loved ones. What unbearable suffering it causes. I did my best to try to make them feel closer together by placing photographs by my patients' beds, playing music and caring for them as best I could.

Meta made it in the end. After ten days she came off the ventilator – we were so relieved when the message came through. She returned to our hospital and spent several weeks at the rehab centre. Two months later I went to visit her at home. She's still far from fully recovered, but she was beaming, and the sight of her did me a world of good.

Something seems to have changed between us. Although Meta had always been very friendly, she still reined in her feelings and kept a professional distance. During her sadness that Saturday afternoon, I reached out and took her hand. I don't know whether she appreciated it, I just did what came to me at the time. When I saw her again, I saw what it had done to her. In my darkest hours, you were there to help me, she said. It feels as though our paths in life crossed, and for a brief time we walked side by side. We now share such a deep experience, one that neither of us will ever forget.'

No answer

Beverley Hunt, haematologist

'My younger brother Philip was the black sheep of the family, and over the years we gradually slipped further and further out of touch. He used to be a brilliant hotel manager, until he became an alcoholic, and we drifted apart. He lived alone in a little town on the south coast of England; it must have been three years since I saw him last.

In early April he contracted the coronavirus, with a fever and coughing. He decided to go into isolation, and every day a good friend of his dropped off meals that he would collect at the front door. One day he didn't answer the doorbell, so his friend got worried and called the police. He was found dead in his apartment, probably from a heart attack.

His death came in what was probably the most stressful period of my career. London was hit hard by the virus, our hospital had closed all its wards and was only accepting COVID patients. The number of intensive-care beds – usually only forty – nearly quintupled, and we were planning to go up to 400 if necessary. Initially we were learning about the illness on our feet; we had so many sick patients, and there was no body of existing evidence to point us in a clear direction. After some time, it became apparent that many patients were affected by the most extraordinary "sticky blood". It was something I'd never seen before. Day after day we saw it happen: the high rates of blood clots causing strokes, deep-vein thrombosis and heart attacks.

So I was getting sucked into all the care, I was really wound up, trying to improve the treatment of patients with thrombosis and COVID not just in our own hospital but nationally and internationally as well. And then I was confronted with the death of my brother, from the very condition that I was researching. It was really difficult to sort my head out. I still feel a lot of guilt because he died on his own, and I am terribly, terribly sad. I sought escape by working like a maniac.

His death impassioned me to work tirelessly. It drove me to keep reading about COVID, to be right on top of all of the literature, so that I can truly understand the origin and development of the disease. I give about three webinars a week, talk to other doctors and patients, and I have also helped to produce the national, and some international, guidance. For me, the whole COVID period has been overshadowed by Philip's death. I feel like I need to do all I can to honour my brother, in my own way and with my own skills, to improve understanding and care for thrombosis and COVID patients, even if it only changes a few people's minds, or the way doctors prescribe medications.

I just worked and worked. I didn't take any time off to deal with my grief, and am still struggling with it. But I can't keep going much longer. I have decided to take a week off, a week when I won't think about COVID at all. I plan to just bake a cake and let the sadness in.

The last time I saw him, years ago now, he seemed very well, but there was always that sense of worry about him. When he drank, he became a totally different person. He was a sweet brother, but everything just went very wrong in his life. We had to distance ourselves from him, for our own protection. All we have now is an overwhelming sense of sorrow about what happened.

Somehow the funeral was wonderful. There were people who said what a rascal he was, but how much they enjoyed his company. We also heard of how naughty he could be; he was a very funny man. I was approached by his friends, who recalled memories of how Philip often spoke about me. He was extremely proud, they said. It made me well up to hear that – I was so touched. What a terrible shame that I never knew it until then.'

Silent impact

Jim Down, intensive-care doctor

'Her father called me, and straight away I could sense the fear in his voice. His nineteen-year-old daughter had been admitted to our hospital, but now that the coronavirus was everywhere, he was terrified of what might happen. She had had a liver transplant the year before, along with countless other operations. She had a colostomy bag and was prone to both infections and dehydration, which can sometimes cause kidney problems. She was incredibly vulnerable, a COVID infection could be fatal, and the hospital – the very place where she ought to feel safe – presented an enormous risk. COVID patients had been streaming in for weeks; we were inundated.

She was placed in a well-protected ward, but still her father became more anxious with each passing day. The COVID-free areas in our hospital were getting smaller and smaller, and she had to be moved several times. It was as though COVID were closing in on her, like an enormous tidal wave that she was powerless to avoid. After ten days they took her home; she wasn't fully recovered, but she was well enough to leave the hospital. She had remained COVID-free, and I breathed a sigh of relief. I know her father personally – he's the brother of a friend of mine, and I'd spoken to him on the phone a few times during her stay. His daughter's illness had caused her such a difficult teenage life, and he wanted to do all he could to protect her.

And then came the telephone call from emergency. It was the

Wednesday before Easter, and I had the night shift and things were hectic. We were almost at the peak of the pandemic. A nineteen-year-old female had been admitted with kidney failure and extremely high potassium levels – a life-threatening condition that can lead to heart failure. I knew instantly that it was her. Under normal circumstances I would have had her taken to intensive care straight away, but that night we were full of COVID patients. There was only one more bed available, but I didn't dare take the risk: under no circumstances did I want to put her somewhere where she could become infected. After talking to the emergency doctors, I decided to leave her where she was, to eliminate any unnecessary risk.

I felt I ought to go and see her, so I walked downstairs. I will remember those few minutes for the rest of my life. It was dark and quiet outside; I walked through a virtually empty hospital and found the emergency department almost abandoned. The contrast with the hectic nature of our department could not have been any greater – I walked past room after empty room, and there was not a soul in sight.

Those few weeks were all about the intensive-care patients. The pandemic had locked me inside a world of unpredictable illness – we were confronted with a stream of incredibly ill patients, and what we saw put us all in shock. When I stepped outside that world for a few moments, it suddenly dawned on me that there was so much more going on.

That night, I felt the silent impact of COVID for the first time. I saw how menacing the coronavirus must have seemed to other patients. The sight of all those empty rooms made me realise how scared they were to come to the hospital. It is an immense tragedy, and we now know that it cost some of those patients their lives.

It was during those few weeks that I assumed I would get COVID. I was surrounded by it, at significant risk, and many of my colleagues had already contracted the disease. It didn't happen to me, but I recognise and understand the fear.

When I arrived at emergency, I kept my distance from her for fear of presenting an infection risk. I just peered through a gap in the curtain. They admitted her to an emergency room and kept a very close eye on her, sparing her a potentially risky journey through the hospital. After a few days she was ready to go home. Her parents were so grateful that they sent me a huge box of wine. I felt a little awkward – what had I done that was so special?

I hardly know her, we've never had a conversation, and yet she has had a huge impact on me. Fear was everywhere: that's what she taught me that night. I can only imagine the anguish it was causing all the other patients beyond my intensive-care unit. I would like to meet her properly one day, and we've decided that's what we'll do . . . when this is all over.'

Anonymous

David Pattyn, anaesthesiologist

'Their eyes are all I can remember, wide-open eyes reflecting their fear of what was to come. I saw them gasping for breath, surrounded by complete strangers dressed in spacesuits, uttering words they couldn't understand from behind plexiglas visors. They were all alone and isolated from their loved ones, to whom they may have wanted to say some final words before disappearing into limbo, a no man's land, unsure of whether they would ever return.

When the first wave of the corona pandemic hit our hospital, we all agreed that the anaesthesiologists would assist with the technical procedures. If a patient deteriorated to the point where they required ventilation, we (the anaesthesiologists) would put them to sleep, intubate them and insert cannulas so that medications could be administered. That's our day-to-day job in theatre, it's what we're good at, and it allowed the intensive-care doctors to focus on caring for the patients in their own departments, which were rapidly becoming overcrowded.

Before an operation, we're used to reassuring patients and explaining calmly what's about to happen to them. But in the months when corona took over our hospital, the human dimension seemed to disappear. There were so many patients, all with the same disease, who we tried to reach with muffled words from behind our plexiglas visors. The intensive-care unit was full of people asleep, often ventilated and lying on their bellies (which hid their faces from us), and all of them totally alone, since family visits weren't allowed due to the infection risk. The COVID patient was an anonymous patient: I remember none of them individually, only some details here and there. And, paradoxically, it's precisely that anonymity that had such a big impact on me.

Shortly before being ventilated, patients were always informed of the associated risks. We told them it was uncertain whether they would survive. Their eyes betrayed a kind of fatalism, as if to say "so this is it". It was a look that touched all of our hearts, and they must have been extremely fearful as we put them to sleep. The first time I had to intubate a COVID patient I wasn't so aware of it; I was mainly shocked by how quickly that man deteriorated before our very eyes. He turned ash-grey within half a minute.

I didn't truly realise what was happening until I brought my camera to work one day. I take nature photographs in my spare time, and it wasn't until the corona crisis that I felt the need to document any of the happenings in our ward. After my shift I photographed some colleagues hooking up a patient to the ventilator. Standing there on the sidelines, looking through a lens, I suddenly understood the seriousness of our communication problems. My colleagues' words were barely intelligible even to me, and they certainly didn't register with the patient either. He was concerned with one thing and one thing only: getting enough oxygen. There was no time to ask how he was feeling, to offer a personal touch.

Effective communication with patients and families – only when it's no longer possible do you realise how essential it is. We did what we could to retain some human contact, but we were unknown and unrecognisable to patients, and they were just as anonymous to us. We could offer no more than mechanical treatments, which left me feeling extremely powerless and distanced. When I took more photos later in intensive care, I saw what a surreal atmosphere had developed there. Never before had things been simultaneously so hectic and yet so still. Only the soft beeping of the IVs and the gentle whirring of the ventilators could be heard in the background. In my photographs, none of the patients are visible; you might say it was symbolic of what was going on.

I intend to collate my photos into an album for my colleagues. In a few years' time, we may have forgotten just how intense this period has been. The pictures will provide evidence of how hard people worked, how focused we were, and of our team spirit. We grew closer together, and now we know that we can depend on one another when it really counts. But what the patients went through was heartbreaking: all alone in the hospital, with no idea whether they would come out dead or alive. And we could do nothing to allay their fears or ease their loneliness. All we could do was promise to take good care of them.'

Stubborn

Mervyn Singer, intensive-care doctor

'She was a little old Italian lady in her early sixties, and for years she had been looking after her chronically ill husband. She was many years his junior and very strong-minded. She didn't care much for conventional Western medicine, instead swearing by the use of herbs and homeopathy. That had almost spelled his demise: when he came down with serious diarrhoea, she administered those herbs of hers for days, and he eventually ended up unconscious in the emergency department. He needed resuscitation and life support, which is how he came to us in intensive care the week before Christmas. She came with him and never left his side.

After several days he came off the ventilator, but his condition worsened. His heart was already in bad shape: he'd had strokes in the past, was wheelchair-bound, suffered from diabetes and now his kidneys were on the verge of collapse. He was nearly eighty; were we really supposed to pull out all the stops to keep him alive? The problem was that we couldn't ask him directly: he was conscious but drowsy. We obviously involved his wife in life-prolonging discussions, and the bizarre months that followed are still very clear in my mind.

Initially we were on the same page: we were to continue our treatment but not escalate to heroic measures. But on Christmas Eve, she suddenly changed her mind. I want everything done to keep him alive, she said to the perplexed junior doctor on duty. I still remember our conversation the next morning: we went round and round like a revolving door. As doctors we must of course avoid pointless medical procedures, but I didn't want to enter a conflict with her either. I explained that it really was better to leave him off the ventilator, that it would be cruel to resuscitate him if his heart failed again and that our job was to act in his best interests. She seemed satisfied. But then we received a telephone call from the police: she had lodged a report accusing us of attempted murder. I'd just finished my shift and handed over to a colleague, who eventually capitulated to her threats. His bed then became the focal point of a battery of machines that were keeping him alive. And while his condition actually improved, his heart had weakened to the point where he could no longer survive without ventilation. And so there he remained in our ward, speechless and powerless, with that woman fussing about him all day long and doling out her forceful, aggravating commands. We had to issue no fewer than three banning orders against her for using obscene and aggressive language. But her husband burst into tears whenever we barred her, so we always let her back in again. All the while she kept lodging complaints – to the police, her MP, the hospital chief executive – accusing us of trying to murder her husband. We did all we could to keep her satisfied – even bringing in a consultant homeopath to pacify her.

What didn't help was that she had terribly bad breath: you could smell it a mile off, and she was there day and night. The nurses on duty always switched shifts with her halfway, since six hours of her hectoring and halitosis was more than anyone could bear.

The situation seemed hopeless, until one day we heard that she had hired a ventilator to take him home. The personnel for operating such specialised equipment would cost hundreds of thousands a year, and social services refused to pay – but she would not be dissuaded. I can still remember sitting down with her in an attempt to talk her out of it. You don't have the proper training to ventilate him, I said. Doesn't matter, she replied, I've seen the nurses do it. Any idiot can do it.

It would spell his death, we all knew it. Could we just let him go like that? We consulted the hospital lawyer, who saw no objections. The patient was of sound mind, we had explained the risks, and if he wanted to leave against our recommendation then that was his decision. And so, shortly before the start of summer and six months after arriving at our intensive-care unit, he trundled off home with his wife and her ventilator. We gave them one day tops. Imagine our surprise when, nine months later, we received a phone call from the ventilator company. She could no longer afford the hire, so what could they do? Surely they couldn't just pull the plug on him? I remember my astonishment after hanging up the phone: she'd actually done it, she'd managed to keep him alive on the machine all that time. Shortly afterwards she moved with him to Scotland, where the healthcare system did cover the costs of ventilation.

We never found out what age he reached, or how the story ended. But that's not the point. We all thought we knew what was best for both of them, but we had it all wrong. And it was she who confronted me with that reality. I still remember the man's happiness when he heard he could finally go home. He wanted to keep on living, and she was there for him in her own inimitable way. She taught me that we should guard against medical paternalism, the notion that only the doctor's opinion counts. Patients and their families will sometimes have other ideas, and occasionally the doctors are just plain wrong. It was a valuable lesson in humility.

Speechless

Nigel Jack, anaesthesiologist

'He was a man of about seventy, and had been lying there in the hospital bed for weeks, staring at the ceiling day in and day out. Paralysed by a severe stroke, life had suddenly ground to a complete halt. He couldn't move or talk, his head just lay motionless on the pillow – even swallowing was impossible, so he received fluids and food through a tube in his nose. But he could still lift his right arm a little, and after several weeks he used that very arm to rip out the feeding tube.

We put it back in straight away, and I can still remember how horrid the procedure was for him. Two days later, it was out again. He must have come to understand the situation he'd ended up in, and knew the tube was keeping him alive. He was clearly using the last of his strength to send us a message. I couldn't blame him: there would be no end to his suffering, there was no hope for any recovery. Despite the nurses' devoted care and attention, his pressure sores grew larger and more painful, so turning and washing him was always a frightful ordeal.

I had just started my first job, as a house officer in a Scottish hospital. Each room had twenty beds, separated by curtains. I shall never forget where his bed was, halfway down on the left side. One day I went to sit next to him and asked whether he realised what would happen if we didn't replace the tube. I saw a slight nodding movement of his head. It means you'll die, I said

– do you understand? Again the slight nod. I then asked him: is that what you want? Again he nodded. I went to the physician and told him about what the patient had been trying to say. He didn't want to go on – surely that was a wish we ought to respect? I was given a firm but friendly warning: as doctors, it is our sole duty to treat patients and to keep them alive. I was deeply frustrated, but I was only a junior doctor and my opinions counted for little.

Then, one morning, a geriatrician came to see him. I told him how things had been over the previous weeks; he looked at the powerless man and suddenly showed sympathy and understanding. He proposed moving the patient across to his own ward, a little further down in the hospital. The physician agreed. I sat at the old man's bedside and explained to him that I couldn't fulfil his wish, but that he would be put under the charge of another doctor. I never saw him again after that. One week later, the geriatrician called me to say that the man had died. After he'd ripped out his feeding tube again, the doctors decided to just leave it be. They gave him sedatives and painkillers, and he died peacefully a few days later.

The events of those first few months in my job have stuck with me throughout my career and left me with the lifelong conviction that being a good doctor is about more than just *treating* people – it's about *helping* people. And helping people also means knowing when not to treat them. That's also what it says in the most recent version of the Hippocratic oath, which states that doctors should apply all measures required for the benefit of the sick, but avoid overtreatment. Do not chase the angel of mercy away from the foot of the bed – that is what I learned from this helpless yet courageous man. He never said a word, and yet he taught me a very important lesson.'

Last words

Karim Brohi, trauma surgeon

'She was seventeen and had been knocked off her bike by a truck. She came into our hospital awake and talking. We figured out that her pelvis was broken, and she was bleeding into her abdomen, which had in turn caused her blood pressure to fall dangerously low. We followed the usual protocol, gave her saline and extra blood, and quickly rushed her to the operating theatre. I was working as a junior doctor in intensive care; it was my job to monitor her airways and work with the consultant anaesthetist to put her to sleep. I sat beside her at the head of the operating table as the whole trauma team bustled around us, getting things ready.

She was anxious and in shock. Not twenty minutes before, she'd happily been on her way somewhere – perhaps to school, or to visit friends – and now she was lying here, fighting for her life. I talked to her and explained what we were going to do. Just before she went off to sleep, she looked at me and asked, "I'll be okay, won't I?" And I said, "Yes, you'll be okay." The surgeons were planning to stabilise the pelvis with a metal frame, but when they made the first incision, it started bleeding and wouldn't stop. Then they opened up her abdomen and massive amounts of blood welled up – she just kept on bleeding. The vascular surgeon took heroic steps to try to stop it; we were pouring new blood and fluids into her body, but then she started bleeding from her mouth, her eyes – well, from everywhere really. Her blood was like cher-ryade: she'd been given so much fluid, all that was coming out was a kind of reddish-coloured clear liquid. She died on the operating table, fewer than forty-five minutes after I'd reassured her everything would be okay.

Three-quarters of an hour later another patient came in. He was very similar, awake and talking, and also died within the hour. Although my recollection of that second patient is quite hazy, I will never forget the first girl. She is very, very clear in my mind. Partly because of her last words, but also the fact that I will always remember how she died right there in front of us, despite the operating team doing their utmost to try to save her.

A few days later, all the hospital specialists came together to discuss in detail what had happened. We'd done things exactly by the book, so why did it all go so horribly wrong?

We launched an investigation, which grew into a personal research project that has been my main interest for many years now. It gradually became clear that our approach, and the approach taken by doctors all around the world, was misguided. Since then we have learned that trauma patients develop major blood-clotting problems. With extensive injuries they form blood clots slowly, and the clots that do form are very fragile and break down rapidly. Without the ability to clot they bleed more, making surgical repairs difficult or impossible. And what were we doing all that time? Pushing lots of fluids into patients to try to raise their blood pressure, without realising that it only made things worse by flushing away the precious few clots that they *could* still produce. Under those conditions, it was no surprise that our girl died on the operating table, bleeding from everywhere.

Things are very different now. Our research has led to major changes in the protocols for the treatment of bleeding trauma patients all around the world. We've switched to a completely new method called "Damage Control Resuscitation", which aims to preserve and improve clotting. We keep their blood pressure low, give them agents to enhance their clotting and have developed special techniques for controlling blood loss early, even on the streets. It's been twenty-four years now, and if that girl were brought into hospital today, I believe she would survive. I'm sure of it.

Most doctors have a sort of internal graveyard of patients they've lost. It's like a constant weight that we carry around with us. As doctors we must try to learn from every death that occurs, even if no mistakes were made. I often think back to that seventeen-year-old girl, for a couple of reasons: she altered the course of my career, and she also led us to discoveries that have helped save the lives of countless patients since.

I don't think I ever knew her name, but she taught the whole world a lesson.